# THE WATCH STANDS AS WRITTEN

*The Watch Stands As Written published by:*
*WILDBLUE PRESS*
*P.O. Box 102440*
*Denver, Colorado 80250*



*ISBN 978-1-970361-36-0 Hardcover*
*ISBN 978-1-970361-37-7 Trade Paperback*
*ISBN 978-1-970361-35-3 eBook*



*Interior Formatting and Book Cover Design by Elijah Toten*
*www.totencreative.com*

# THE WATCH STANDS AS WRITTEN

*Voices from the Midwatch*

DAVID E. JOHNSON

GARY M. GUINN

KENNETH D. KEITH

WILDBLUE PRESS

WildBluePress.com

# CONTENTS

Acknowledgments 7

Preface 9

Introduction 13
*Sidelight 1: Land, Sea, and Space: Unusual Places for Poems* *21*

Chapter 1 - As Rhyme Goes By: Complaints About Writing Poetry 23
*Sidelight 2: Midwatch Misery: Complaints About Writing Poetry on New Year's Day* *37*

Chapter 2 - Triumph and Tragedy: The Space Race, Popular Culture, and Conflict 41
*Sidelight 3: Should old acquaintance be forgot...: A Midwatch Poet Responds* *56*

Chapter 3 - Wine, women, and …… 61
*Sidelight 4: Ships as Women: Personal Pronouns She/Her* *79*

Chapter 4 - Late Night Laughter: Humor on the Midwatch 84
*Sidelight 5: Lost, but not Forgotten* *97*

Chapter 5 - Duck and Cover: The Threat of Annihilation in Cold War Deck Log Poems 99
*Sidelight 6: Give Me Liberty……* *108*

Chapter 6 - To Sea or Not to Sea? That's a Good Question 113
*Sidelight 7: A Mind of Metal and Grace: Charles J. McMahon, Jr.* *131*

Chapter 7 - Phantoms of the Deep: Hauntings and Visitations at Sea 134
*Sidelight 8: Admiral John Hoskins: The Man with a Wooden Leg and an Iron Will* *144*

Chapter 8 - Leatherneck Bards: Marines Embrace Unusual Duty Aboard U.S. Navy Ships 146
*Sidelight 9: Women in the Navy*

Chapter 9 - Midwatch Mischief: Navigating Unusual Rhymes in Deck Log Poetry 158
*Sidelight 10: The Lone Haiku: Yasuto Tana* *176*

Chapter 10 - Midwatch Deck Log Poetry Aboard US Submarines: The Early Cold War Years 179
*Sidelight 11: The Long and Short of it* *198*

Chapter 11 - Graphic Embellishments: How Design Added Beauty to Deck Log Poems 201
*Sidelight 12: From Navy Steel to Kinetic Sculpture: James L. Seawright, Jr.* *215*

Chapter 12 - Bards of Viet Nam 218
*Sidelight 13: A Navy Man for Life: Wayne Philo Hughes, Jr* *236*

Chapter 13 - Postscripts 239
*Sidelight 14: This, That, and De Otter: Playing with Language on New Year's Eve* *249*

Epilogue 252

Source Notes 260

*Appendix A - Oh, the Nights are Calm and the Breeze a Balm. . . Excellent Poems From the Midwatch: A Sampler* *271*

# ACKNOWLEDGEMENTS

We are grateful to the folks in the Catalog Division at the National Archives for helping us more easily locate January logs by developing a search algorithm for their vast collection.

Thanks to Rachel Decker for sharing her knowledge of illumination and calligraphy that assisted us in using proper terminology in Chapter 11 about graphic embellishments in deck logs.

To our spouses, Simone, Mary Ann, and Connie, thank you for all your support and encouragement as we tackled this project.

# PREFACE

The impetus for this book began more than ten years before its publication. While researching the experience of his father-in-law aboard USS *Patterson* in WWII, Dave Johnson found poems entered into ship logs for the midwatch (midnight to 4am) on New Year's Eve. This departure from the usual no-nonsense content in ship logs was surprising. Further research into this unusual US Navy tradition led to a series of blog posts titled *Midwatch in Verse.* [1]

Gary Guinn joined Dave in compiling the first book-length treatment of the midwatch poetry tradition - *Midwatch in Verse: New Year's Deck Log Poetry of the United States Navy, 1941–1946.* [2] The book tells the stories of 29 ships, 45 poems, and the authors of the poems. To promote awareness and knowledge of this curious tradition, Dave developed a website – *The Midwatch in Verse Project* (http://www.midwatch-in-verse.com).

As greater numbers of post-WWII ship logs became available for review online from the National Archives, the authors asked Ken Keith to join them in an analysis of poems appearing in deck logs from 1947 through 1978. The content of this book came from over 10,000 deck log poems taken from over 16,000 logs. From the analysis of this voluminous database of poems, several themes emerged that reveal the zeitgeist of the nation and the individual sailors who wrote the poems—their aspirations, their fears, their engagement with the critical events of the era. This engagement is especially true during the Vietnam War but also reflects a broad range of socio-political issues/conflicts at other times and the immediate and personal concerns of young officers serving their country far from home and family. These themes and others guided the organization of this book

The deck log poems were downloaded from digitized logs maintained by The U.S. National Archives. Most of the logs downloaded were from the years 1956 through 1970, although there were several hundred logs from the years 1971 through 1978. The digitizing of these logs was made possible by funds provided by entities such as the Veterans Administration, which uses the logs to verify ship activities and locations when veterans submit claims.

The authors developed their material with the following historical context in mind:

After WWII, the US Navy experienced an unprecedented reduction in the size of its fleets, mirroring its growth during the war years. Active ship force levels dropped from nearly 7,000 at the end of the war to about 1,200 just a year later.[3] The Navy had to contend with two wars, Korea and Vietnam, neither of which bore much resemblance to WWII with respect to force levels and tactics. The total war model of WWII was replaced by "limited" conflict. And the Cold War developed, putting pressure on the Navy to keep the Soviet-aligned countries in check as China began its rise to become another potential adversary. Middle Eastern tensions prompted a consistent naval presence in the Mediterranean Sea. The Navy also participated in diplomatic efforts to assist US allies and to offer aid to those countries that might become allies. The poems included in this book span the aforementioned wars, the Cuban Missile Crisis, and the period of détente.

The book focuses not just on the poetry itself, but also on what the poems reveal about the young Naval officers and the country at large. Excerpts from the poems support general assertions about thematic material, and a selection of complete or nearly complete poems appears in Appendix A, giving readers a better sense of the material in its original form. In addition, sidelights here and there throughout the book illustrate aspects of the human stories behind the experience—among them, poets pleased to be reminded of their long-ago verses, the only log written by a woman, examples of unique poetic style, and young sailors who went on to distinguished Naval careers.

Here, then, are poems that reflect the mood of a nation, through the eyes of those who sacrificed on its behalf. These are their stories, told in the voices of those who stood the midwatch.

# INTRODUCTION

It's January 1, 1926, aboard USS *Idaho* (BB 42) anchored in San Pedro Harbor, California. A young ensign, E. V. Dockweiler, is serving as Officer of the Deck (OOD) and is responsible for compiling the ship's deck log for the Midwatch – midnight to 4:00am. Deck logs contain a chronological record of the ship and include information about the ship's condition, its location, movements, and any other significant information about events surrounding the ship. Deck logs often contain information that, to the untrained eye, may appear uninterpretable with a plethora of compass headings, military acronyms, and maritime jargon. (See Appendix X for additional information on deck logs)

Ensign Dockweiler knows, though, that a US Navy tradition allows him to deviate from "just the facts" and to apply some creativity to his log entry by writing it in verse. So early in the log he writes,

> We are anchored in Pedro harbor
> Tho there isn't much of a lee
> And why they call it a harbor
> Is something I never could see

He follows this verse with information about *Idaho's* current conditions, all in rhyme, and finishes with,

That's all the dope this morning
Except, just between us two
If the Captain ever sees this log
My gawd what will he do?

Capt. Arthur St. Clair Smith signed off on the log, adding this note: "The Captain is glad to see that the old Navy custom of writing up the first watch of the year in rhyme is known to the younger members of the Service. The watch stands as written."

Clearly, Captain Smith knew of the tradition, yet this 1926 vintage poem remains the oldest midwatch New Year's Day poem to be identified. The Captain's comment implies that the tradition's origin predates Ensign Dockweiler's poem, but no earlier deck log poem has been found. Hundreds of logs dating back to the mid-19th century have been checked, but none yield a poem. However, although specific origins of this tradition remain unknown, there are some clues suggesting the idea, or at least the glimmer of an idea, may have arisen in the 19th century.

On January 1, 1880, the USS Jeannette was trapped in Arctic ice. Nevertheless, the midwatch log noted the celebration of the new year[1], and the skipper's journal reported that the ship's meteorologist wrote rhymes featuring members of the crew[2]. Logs further acknowledged New Year's Day and the celebration of its arrival, without poetry, during the next few years. Then, in 1894, a group of Naval Academy midshipmen, having written poems while on an Atlantic cruise aboard USS *Constellation,* determined to create a yearbook[3]. That yearbook, which would become the world's largest, and continues to be published at the Naval Academy, was the *Lucky Bag.*

*The Lucky Bag* title was no doubt borrowed from naval tradition. The lucky bag was a repository for lost or misplaced items found aboard British naval ships, and the term first appeared in 1788 in the writings of James Hurdis, an English clergyman and poet.[4] In the U.S. Navy it had a history dating from at least a half century before Lucky Bag came into being at the Naval Academy. In the 1840s, Commodore Matthew F. Maury wrote a series of magazine articles under the title "Scraps from the Lucky Bag," and even penned a poem, a parody enumerating the kinds of articles that might be found in the lucky bag[5]:

> Shoe of waister and middy's sock,
> Wing of soldier and idler's frock,
> Purser's slops and topman's hat,
> Boatswain's call and colt and cat,
> Belt that on the berthdeck lay,
> In the Lucky Bag find their way,
> Gaiter, sock, and red pompoon,
> Sailor's pan, his pot and spoon,
> Shirt of cook and trowser's duck,
> Kid and can and 'doctor's truck',
> And all that's lost and found on board
> In the Lucky Bag's always stored.

Although neither Maury's poem, nor those of the midshipmen who established the *Lucky Bag*, appeared in a New Year ship's log, they may well have contributed to the beginnings of the tradition. The inaugural issue of Lucky Bag included thirty-five poems, and poems continued to appear in subsequent editions. It is not hard to imagine that some of those early middie poets may have gone on to serve on ships where they could certainly have lent their literary skill to the midwatch verse tradition.

Whatever their origin, midwatch deck log poems, allowed only on the first midwatch of the year, continued to appear in US Navy logs through the 1940s, including on ships engaged in combat in WWII[1]. During the 1950s the tradition flourished, and deck log poems became the rule rather than the exception throughout the 1960s and 1970s. The poems served as vehicles to express a wide variety of emotions on the part of the young officers who wrote them, revealing their humanity and their "American spirit." While the world around them began to splinter, young sailors wrote poems not just about international issues but also about the issues that young men face everywhere.

Being assigned the duty on New Year's Eve was not a popular turn of events for these young men. While their shipmates were partying, the duty officers were spending four long and boring hours, sometimes suffering inclement weather. Complaints were common in the poems.

Ensign F. E. Clark, aboard USS *Claude Jones* (DE 1033) in 1962, complained that he wasn't up to the task.

> Before I attempt this bit of rhyme,
> Let me say I "ain't" no poet;
> And before we're through with these few lines
> You won't need a Degree to know it.
> USS *Claude Jones* (DE 1033) 1962

There were complaints about being "required" to write a poem, often blaming tradition for "demanding" it be done. Perhaps the most common complaint of all concerned the absence of alcohol and female companionship. The duty officer was painfully aware that his shipmates on shore had plenty of both. As Lt. j.g. G. R. Hallberg of USS *Boston* (CAG 1) put it,

> The various services we receive from the pier
> Do not, alas, include wine, women, or beer,

While the deck log poets were complaining, usually with their tongue firmly planted in their cheek, they found novel topics to address, such as ship visitations and hauntings. In addition, they attempted a variety of methods to increase the appeal of their poem to the reader. Among those were the imitation of classic or popular poems, such "Gunga Din," "The Raven," and "A Visit From St. Nicholas" ("The Night before Christmas"); the use of unusual rhyming words; the creation of novel poetic structures; and art in the form of marginalia. And through it all, the poems reflected the dogged humor of the American sailor.

The times in which they lived, however, could not be ignored, and world events occupied the minds of young officers. After the turmoil of WWII, world peace was soon threatened again by a global power struggle between East and West. The rise of a new vison of war, "cold" war, threatened widespread destruction as nations other than the United States developed nuclear power. At the beginning of a new decade, China flexed its muscles, coming to the aid of the North in the Korean War. One result of war and the threat of war was frequent reflection in the deck log poems on the sacrifices of sailors that came before. In 1955, Ensign R. H. Lynch, aboard USS *Taconic* (AGS 17), less than two years after the end of the Korean War, worried about the possibility that more of his fellow sailors would suffer the same fate as those who paid the ultimate sacrifice.

> Too often the world has suffered this pain
> Too often in the sea brave men are lain
> Never to return to their loved ones at home
> But destined the ocean depths to roam.

Six years later, in 1961, Lt. J. L. Brown aboard USS *Queenfish* (SS 393) expressed similar concerns.

> The boys who sailed the Pig Boats in the days of World War II
> Could tell us all of sacrifices such as we never knew

The lads that sailed in Gudgeon, in Growler, and in Scamp
Now here beneath these very waves they make eternal camp

The new world order pitting the Soviet Union, and eventually China, against the U.S. permeated military doctrine during these decades, and references to it appeared frequently in US Navy deck log poetry. Being at the vanguard of the overt and covert machinations of Communist countries, sailors often expressed their fears of growing threats and hopes for a future of freedom for all. One of the more eloquent and strident midwatch poems appears in USS *Goodrich's* (DDR 831) New Year's log in 1957. The poem expresses feelings of frustration at not being able to help those perceived to need it desperately. In part, it may refer to the helplessness at not being able to aid Hungarians who just months before suffered a jack-booted crushing of their uprising against the Soviets.

Dark were the clouds, this past year.
And pleas for help, we did not hear.
Stormy the seas, so bloody and Red.
We were afraid to help, so they are dead.
Our conscience hurt and we did blush
We said hide in my house, it is so plush.

… … … …

Theres a rising tide so bloody and Red.
The time is near to choose our head.
Oily words will calm it naught
This is a storm that must be fought.

… … … … …

When again is heard a cry of woe,
Out to sea our might will go.

This New Year we must go forward,
Last year I felt, so like a coward.

R. A. Turner

R. B. Stephenson, RDC, USN

The poem expresses some optimism for the future where freedom reigns, but overall the words reflect a pessimistic view of world affairs. The rumblings of another war were already testing America's determination. In a few short years, the US, already giving aid to the French in Vietnam, would be embroiled in a long and bitter conflict there. Beginning in the 1960s, and continuing through the Cold War years, the dark cloud of the threat of nuclear war hung over the country. The mushroom cloud, nuclear bomb shelters, and the phrase "duck and cover" became part of the American lexicon.

Because the Cold War was global in scale, the U.S. and the U.S.S.R. attempted to curry favor with many countries to insure positive relations and alliances. The U.S. Navy endeavored to do this through positive interactions at ports of call and through humanitarian assistance. Lt. j.g. H. R. Hoyt, aboard USS *Canberra* (CAG 2), ends his 1961 midwatch poem with some reflection of a recent cruise.

The whole world is searching for friendship and peace
And our job ensures that that search will not cease.

. . . . .

Our world cruise has ended. Our search is not through
We have found goodwill – in the hearts of our crew
Who've spread that goodwill round the world – far and near

> In hopes we'll have peace, this, and each coming year.

But as Bob Dyan proclaimed in his immensely popular song in 1964, "The Times They [Were] a Changing." For the modern world, the rift between East and West would widen, and Communism would become the ideological force that stoked American fear and against which the US would pit all its energy.

For the Navy, modernization would include widening participation by women in the active fleet. The Cold War years were marked by turmoil and significant social change both in the US and throughout the world. But through the years, young officers on the New Year's midwatch continued to take up the curious, extraordinary tradition of writing a midwatch in verse.

# SIDELIGHT 1

## Land, Sea, and Space: Unusual Places for Poems

In addition to the midwatch on January 1, poems may appear in logs at other times. It is acceptable to enter the first watch after a ship's commissioning in verse.[1] These poems appear infrequently because commissioning is a one-time event for each ship (Note: A few ships experience decommissioning only to be recommissioned later when the ship is needed again). It is not uncommon for the poem to be written or co-authored by the commanding officer. Officers of the deck (OODs) mostly write the poems themselves, but sometimes they are assisted by other crew members, often with attribution. Occasionally, someone other than the OOD of record writes the poem. Again, the log may pay appropriate attribution to the actual author

Poems in New Year's Day logs do not have to come from ships. Shore-based units such as a Naval Air Station (NAS) or a Navy hospital on occasion fill their first log entry of the year with metric refrain.

Midwatch poems may also occur off world. Cmdr. Bill Shepherd, a Navy captain, posted the following poem[2] while aboard the International Space Station in the ship's log for 0000 01 Jan 2001:

> We sail onboard space station "Alpha"
> Orbiting high above Earth, still in night
> Traveling our destined journey
> beyond realm of sea voyage or flight
> A first New Year is upon us
> Eight strikes on the bell now as one
> The globe spins below on its motion
> Counting the last thousand years done.
>
> .....
>
> On this ship's deck sits no helm now
> Rudder, sheet and rig long since gone
> But here still—a pull to go places
> Beyond lines where sky meets the dawn
> Though star trackers mark Altair and Vega
> Same as mariners eyed long ago
> We are still as wayfinders of knowledge
> Seeking new things that mankind shall know.

# CHAPTER 1

## As Rhyme Goes By: Complaints About Writing Poetry

Tis hard enough these words to weave
On every duty night
But when it comes to New Year's Eve
This curse in verse we write.

Thus, Ensign R. W. Lewis aboard USS *Midway* (CVA 41) in 1965 participated in one of the most consistent charades in the history of midwatch deck logs—the mock complaint describing the tribulation of having to write the first midwatch of the year in verse. Lewis adds, with a touch of self-censorship,

We're bound by duty, per Navy regs,
To give our hard earned time.
Yet this question for an answer begs
Why must this ______ thing rhyme?

The chorus of complaints among deck log poets pretends at times to reach the level of weeping and wailing. However, the common note in all complaints is humor, a poem written with keen wit as the poet deftly performs a task that he asserts is impossible.

## Not a Poet, Not a Bard

Deck log poets complain that the assignment is unfair, and the poor officer of the deck is unequipped to perform his duty.

> Me write the midwatch? I thought it was jest
> My English is poor, I have no finesse
> I'd wander and falter and end in distress
> Surely they wouldn't put me to the test.

So began the midwatch log on USS *Nahant* (AN 83), a dual mission salvage ship and net tender, in 1958 in Little Creek, VA. The Officer of the Deck, Ensign Dale M. Blount, expressed a sentiment that was common among the young officers who were saddled with the first watch of the year and faced the daunting task of writing the log as a poem. Ensign Blount's initial surprise was followed by acceptance of the cruel reality.

> . . . . there was my name on the midwatch line
> It's true I saw and the task is mine
> To write the mid watch and write it in rhyme.

Lt. j.g. (junior grade) C. E. Goodwill, aboard USS *Brownson* (DD 868), resting on keel blocks at Boston Naval Shipyard in 1961, discovered that the task was clearly more difficult than he had anticipated. He asserted that writing a deck log any other time was easy because it was a straightforward expression of facts written in a well-known form.

> But this first log must be correct,
> Complete and in addition,
> It must be written in rhymed verse
> By ancient sea tradition.

Though Goodwill said he initially thought it might simply have been an enjoyable chance to be creative for a change, his actual experience differed markedly:

> . . . now I've stood, pen poised in hand
> For better than an hour
> But no ideas, no rhymes I find,
> Verse seems beyond my power.

But the truth is that the hundreds of similar complaints in the poems by young officers on New Year's Eve belied a clear reality—though many lamented "I'm no poet," those who chose to write in verse waxed quite eloquently in their lamentations. The lament of Senior Chief Petty Officer J. L. Skinner, aboard USS *Prairie* (AD 15) in 1970 at Subic Bay, seems apropos of the fact that *Prairie* was an old ship, laid down in 1938.

> For many years
> I was no poet.
> Times have not changed,
> For the above verses show it.

Ensign E. A. Selby, at Long Beach Naval Shipyard, expressed the common complaint about having trouble getting started writing a poem.

> The log lies before me,
> A blank page is showing;
> I'm just not a poet;
> I can't get this going.

USS *Lyman K. Swenson* (DD 729), 1958

But Ensign Selby must have enjoyed the work once it was begun. His log was one of those occasional poems in which even the signature rhymed:

> E. A. SELBY – NEWLY AN ENSIGN

U. S. NAVY – THE BOOT ON THE SWENSON.
[BOOT is lowest ranking officer]

In 1961, Lt. R. C. A. Sullivan, writing aboard the submarine USS *Redfin* (SSR 272), which a little over a year earlier had become a laboratory and training ship for the testing of inertial guidance systems used in Polaris submarines, expressed the typical desire for efficiency when he ended with,

No sense wasting ink
Writing verse that doth stink
When I could be "Moored as before."

**A Cry for Help**

Many young officers writing deck log poems complained specifically about how difficult it was to make rhymes while considering the required content of the log. Some poets sought help from the crew or borrowed from other New Year's log poems. Lt. j.g.. H. C. Atwood, Jr., confessed

I'm not a poet, I'm not a bard,
So writing this log was rather hard;
"Moored as Before" is usually enough,
But not for this log, that's why it's so tough.
I've asked for assistance from all of the crew,
(I've stolen a little from the log of the NEW),
I'm sober, disgusted, tired and frost bitten
But above is the info that had to be written.

USS *Holder* (DDE 819) 1956

Aboard US submarine *Remora* (SS 487), a ship with a long history of service (commissioned 1946), Lt. j.g.. K. N.

Meneke, on a 1960 passage from Kobe to Yokosuka, Japan, also reached out to shipmates for help with his poem.

When told tonight "Write the midwatch in verse"
I got so mad I could almost curse
I racked my brain to find a solution
When finally I reached the only conclusion
So now I must give a hearty vote of thanks
To Quartermaster Dick and Chief Radioman Baker
For without the assistance of these fine jongleurs
This log would have been a real back-breaker.

Other New Year's poets called on the ancient gods rather than their shipmates for help creating their poems. There was, of course, a strong preference for Neptune, god of the sea.

Lt. j.g.. D. A. Detweele saw *Neptune* as his motivation for writing the deck log poem.

It really isn't easy,
And it always makes me queasy,
To sit down and write a New Years tale.
But every New Year I seem bound
To write something quite profound,
And may Neptune flog me if I fail.

USS *Van Voorhis* (DE 1028) 1967

Still other midwatch poets bemoaned the fact that, having the duty on New Year's Eve, the traditional celebratory catalysts for inspiration were unavailable. In Little Creek, VA, Chief Petty Officer R. L. Nolan calls on the sea as his *second choice* of muse.

Hail Sea Muse! Be with me tonight,
For I to paper a poem must write.
I have not the help of Bourbon and Gin—

Only my trusty ballpoint pen.

USS *Hermitage* (LSD 34) 1965.

Nolan adds several lines within the poem echoing his fear of failure. For example, at the end of stanza three, he notes parenthetically,

(Just a few more lines and we'll both get some rest
Stay with me, O Muse, you're not at your best).

**Write It Yourself**

It is commonplace in American culture for a person frustrated by a difficult task to say in exasperation, "Do it yourself then." In their mock frustration with their poetry assignment, some young officers challenge the reader to do just that at the end of their poem. Lt. j.g.. B. H. Stewart, aboard USS *Iwo Jima* (LPH 2*)*, an amphibious assault ship with a contingent of Marines practicing amphibious and landing raids off Taiwan and Okinawa in prep for duty in Vietnam, ends his poem with the following:

Now this is the close of an old year and the start of a new day,
And in closing this watch I have something to say:
This is my third year to try this rhyme,
And if you don't like it, ship over and do it yourself next time.

And the bard aboard USS *Pillsbury* (DE 133), Lt. j.g.. R. M. Beers, echoes the same sentiment.

Now if this isn't the best poetry money can buy,
Next year you're welcome to give it a try.

## Tradition Made Me Do It

One of the most common complaints from the Officer of the Deck on New Year's Eve consisted of blaming the long tradition for "requiring" him to attempt a poem in the log. Lt. S. N. Darby, aboard USS *Dace* (SSN 607), a ship sponsored and christened by Betty Ford, wife of congressman and future president Gerald Ford, blames "old Navy custom."

> Now old Navy custom
> Tried and proven by time,
> Says the New Year's log
> Shall be written in rhyme.
> To some it's easy
> To whip out a verse;
> To me (I'll admit)
> It's quite the reverse.
>
> (But what the hell,
> On with the ditty . . .
> It just has to rhyme
> And needn't be pretty.)

On New Year's Day 1968, Chief Petty Officer R. B. MacGregor regrets that "tradition demands" a poetic entry.

> The night is dark; the wind is calm;
> A misty rain is falling.
> It is New Years Day of '68
> As you can see I'm stalling.
> A poetic entry to the log
> Is what tradition demands;
> But unaccustomed as I am,
> I write with shaking hands.
>
> USS *Holland* (AS 32)

Lt. j.g.. D. L. Graham complains about the requirement of tradition but decides on a rather novel response—contacting his congressman.

> The old year's fading fast away, they're ringing in the new,
> And as the first watch stander, I'm really in a stew.
> For in keeping with tradition, I must write my log in rhyme,
> I think I'll see my congressman about drawing overtime.
>
> USS *Henley* (DD 762)

Lt. j.g.. R. T. Schubert, on USS *Rock* (SS 274), while underway independently in 1957, offers a wry twist to the demand of tradition. He uses the fact that he is a Reserve officer to justify *not* following tradition.

> I'm a short time reserve,
> And so proudly I serve,
> But this New Years I'll break one tradition.
> Right now I'm more tired
> Than I am inspired,
> So the log's in the same old condition______

Schubert follows this short poem with a standard prose deck log offering the required information.

### Forgive Me, Father, For I Have Penned

Most young officers of the Cold War era brought to their ships a top-notch education, many having graduated from the U. S. Naval Academy. So it is not surprising that the midwatch poets often allude to the literary canon and the famous poets found there. Some writers felt the need to apologize to their great forebears. In Naples, Lt. John S. McCabe, aboard USS *Topeka* (CLG 8), while wrapping up a Mediterranean tour, expressed concern about how Milton and Shakespeare might respond to his poem.

> Three hundred and sixty four nights of the year
> Mid-watch standing duties are a thing dull and drear.
> But the New Year's Eve watch offers chances galore,
> To prevent these four hours from becoming a bore.
> For "Old Navy Custom" with passage of time,
> Has decreed that I MUST write my entries in rhyme.
> And though Milton and Shakespeare may groan in their rest,
> At my feeble attempts; I can just try my best.

Senior Chief Petty Officer W. C. [name illegible] (MLCS), while managing meter and rhyme scheme very competently, even using interior rhyme in longer lines, still warns against expecting too much from his pathetic attempt.

> Four hours aren't much time in, to create proper rhyme in,
> But the task must be done, that I know,
> So while whistles are blowing and the New Year gets going,
> As Australians say, 'Give her a go.'
> I'm more athletic, and my meter's pathetic
> But I must say that I'll do my best
> And though this poem lacks, the proper syntax
> After all I'm no Edgar A. Guest.
>
> USS *Fulton* (AS 11) 1965.

The turn of the new year 1969 on ocean mine sweeper USS *Venture* (MSO 496), docked at the Mine Defense Laboratory in Panama City, Florida, offered Ensign R. A. Morrow an opportunity to describe the effect his poem might have on Shakespeare.

This poem, I fear did not behave
Set Shakespeare spinning in his grave
To everyone the best New Year
(Where IS my relief? I'd like a beer!)

Some young officers gave a nod to the canonical poets by imitating a particular form the famous poets were known for. This kind of imitation was not uncommon in New Year's deck logs in general, and those who wrote complaints followed suit. The most common of these imitations was in honor of Edward Lear, originator of the humble limerick. Ensign R. P. Onorati did a very competent job with his.

My poetry is really a crime,
But Navy custom requires a rhyme,
So with vigor and zest
I embark on this test,
For this midwatch I've nothing but time.

USS *Charles S. Perry* (DD 697) 1962

Lt. j.g.. E. F. Cole penned another nice limerick while USS *Harry E. Hubbard* (DD 748) was docked in Long Beach in 1963.

We're sad to say on this guard
That among us there's nary a bard,
But we'll write and erase
With a grimacing face
'Til the log we've poetically marred.

None of the deck log poets had the temerity to tell Shakespeare to move over and make room for a new giant of literature. The Naval bards of New Year's Eve were more likely to confess that the great writers of the literary canon had nothing to fear from their own deck-log poems. Lt. j.g.. Robert P. Luce confessed as much at Naval Base Charleston aboard the first of the Guppy class submarines.[1]

Looking back over these lines we call verse
We cannot help think it could be worse
As for poetry, Longfellow have no fear
We just want to wish everyone a happy New Year.

USS *Odax* (SS 484) 1961

Then Lt. G. W. Werenskjold, aboard submarine USS *Tench* (SS 417) in 1965, with a wink and a nod, humbly discredits his own skill, claiming that if he were only Shakespeare, he could write his poem in iambic meter, a simple two-beat form that Shakespeare used often. However, the lieutenant's humble claim is written using a less common and more difficult anapestic meter.[2]

The New Year's log is a Navy tradition
And a poem it must be is the gruesome condition
If only a Shakespeare I were—"There's the rub"
Then this rhyme with iambics I would gracefully dub.

### I Am Not E-quip-ed

One of the most popular ways for Officers of the Deck to complain about the writing assignment on New Year's Eve consisted of a quick two-line quip to close the poem. This structure gives the reader the feeling that the poet is taking a last jab at the tradition. Some of the quips echo the earlier theme of "I'm no poet."

While on Yankee Station in the Gulf of Tonkin in 1968, Lt. j.g.. K. T. Williams apologized for his poem.

In closing I apologize for this trite little ditty
But of poetry I know little, and that is a pity.

USS *Gurke* (DD 783)

Lt. j.g.. N. M. Mays, while anchored in Iwakuni Harbor, Japan, in 1962, apologized more indirectly.

> This entry is closed and signed by me,
> Who never a poet, should try to be.
>
> USS *San Joaquin County* (LST 1122)

In a similar vein, Ensign J. McCorkle in 1958 takes his apology to the extreme.

> And with that, I conclude this little verse
> Most are better, but none could be worse.
>
> USS *Wilkinson* (DL 5)

After struggling to write the midwatch poem, some poets promise to stick with prose in the future. Lt. j.g.. W. G. Loy says,

> This ends the rhyme I was assigned to compose
> And never again will I vary from prose.
>
> USS *Tanner* (AGS 15) 1961

Lt. j.g.. R. G. Yapp, Jr., OOD on the attack cargo ship USS *Andromeda* (AKA 15) in 1955, was clearly disgruntled by the assignment.

> All the ship's services are received from the dock;
> And with this profound saying I'll end this crock.

When more than ten thousand poets attempt to meet the same basic challenge, a deck log in verse, a reader might expect some repetition or borrowing, but sprinkled among the closing quips in deck log poetry, a particular couplet appears repeatedly:

> Well, it's over, it's done—Navigator please note:
> If the log doesn't rhyme, at least it was wrote.

The likely source of the lines was an article published in *All Hands* in December of 1954, which contained the log of USS *Fechteler* (DD 870), written by Lt. R. O. Pyle. But the use of these lines seems to have taken on a life of its own, appearing in deck log poems spread over decades with only slight variations. The example below, by 1st Lt. C. R. Keith, USMC, from the storied Aircraft Carrier USS *Ticonderoga* (CVA 14) in 1962, changes only the punctuation, omitting the colon used in the original.

> Well, it's over, it's done—Navigator please note
> If the log doesn't rhyme, at least it was wrote.

Though the majority of these lines are addressed to the navigator, occasional exceptions occur. On the ocean fleet tug USS *Cree* (ATF 84) in 1955, while resting at Pearl Harbor, Ensign R. R. Moody decided to skip over the navigator and go straight to the exec.

> Well, it's over, it's done—Exec please note
> If the log doesn't rhyme, at least it was wrote.

This particular pair of lines, with their snappy rhythm and humorous wordplay, clearly caught the fancy of many young officers who, through their mimicry, paid homage to their deck log poet predecessors.

## Conclusion

The rich history of US Naval officers turning their deck log pens to the art of poetry and pretending to complain about the "requirement" is a legacy worth preserving. They expressed their chagrin at the difficulty of writing a mundane regulatory log in verse, the "coercion" foisted on them by tradition, the pressure of measuring up to the literary canon. They sought help from any quarter—fellow sailors, the muses, and the gods—and they borrowed and stole what they could from earlier log entries. In the end, they pretended to throw up their hands in frustration, yet some of the most memorable lines in the deck log poems come in their mock complaints.

A literary gem this log is not
Nor a magnum opus in it can you spot
Old soldiers grow older refusing to die
Young ENSIGNS write poetry when asked to comply.

Ensign T. P. Robinson, USS
*Basilone* (DDE 824) 1957

# SIDELIGHT 2

## Midwatch Misery

Complaints about writing poetry while standing the midwatch appear in large numbers of deck logs. Most sailors, like most people, do not fancy themselves as poets. They likely agree with poet Marianne Moore's opening lines of her poem simply titled "Poetry."

> I too, dislike it: there are things that are important
> beyond all this fiddle.
> Reading it, however, with a perfect contempt for it,
> one discovers that there is in
> it after all, a place for the genuine.

But it is not just feeling pressure to write a poem that sailors object to on New Year's Eve. Some of their angst emerges from merely having the duty on a night when celebration is the norm. As Lt. j.g. R. W. Flynn explains in his 1961 log aboard USS *Coney* (DDE 508):

> For all the sad words of prose or rhyme
> The saddest are these, "It's mid-watch time"
> With dreams interrupted, swing out of your bed
> A quick cup of coffee; a trip to the head

Then bored for four hours, you'll freeze in the cold.
While happy civilians ring out the old
And drink in the new with carefree elations.
They've no cause for worry; the fleet is on station

Lt. j.g. P. T. Smith describes the perceived privations of the OOD.

Moored to Pier 7, starboard side
All lines are doubled and securely tied
I'm sitting on my can in the wardroom now
With no party to go to, no liquor, no chow.

USS *Fairview* (PCER 850), 1968

In a similar vein, Lt. j.g. J. A. Zachman writes:

I guess this is it
What a heck of a blow.
New Year's Eve with the duty
Me, a bachelor, with dough.

USS *Agerholm* (DD 826), 1959

What's worse than having the midwatch on New Year's Eve? It's probably having the midwatch on New Year's Eve two years in a row.

I'll sing to you a song of woe,
Now heave ho, me hearties-
I'm twitching on the quarterdeck
While others twist at parties.
But that's not all that's sad to tell,
My hearty boys, heave ho-
This New Year's Eve aboard the WASP's
My second in a row.

F. A. Decker, Lt. j.g.
*Wasp* (CVA 18), 1962

This OOD begs not to be on the watch list again in the upcoming year.

The log is now written, how about a reprieve?
Don't give me the duty, next New Year's Eve!
I'm regular Navy, a Lieutenant J. G.
That I can't write poetry, is easy to see.
I'll sign my name now, and I'm ready to run
The Duty Officer has been

Edward D. Dunn

USS *Caiman* (SS 323), 1964

Some OODs bemoan the deprivations of having the New Year's Eve midwatch and also take advantage of their ability to express their displeasure in ways that would not be allowed on any other watch.

On some gay street, Honcho First Fleet,
The Senior Officer Afloat
Is laughing, weaving, New Year's Eve-ing
While here I sit, the duty goat.

With parties going, gay horns blowing
Here sits sunken, gloomy me
A thirst for wine, in life's full prime
Writing third-rate poetry.

But.....

It could be worse – that thought perverse
Through my vale of tears is seen
It could be worse without this verse that I can write
to vent my spleen

R. Jacobs, Lt.

USS Permit (SSN 594), 1966

Other OODs express themselves in more explicit terms!

Bringing in the New Year, was my one big dream.
But that dream was shattered, when I saw the watch-
list.
Believe me here is one Chief that really is pis__d.

William J. Smith, BMC

USS *Calhoun County* (LST 519), 1962

Lt. W. E. Norton upon being relieved finishes his log entry with a parting shot at this undesirable duty.

My log has been writ, the saying's been said
To hell with mid-watches, I'm headed for bed.

USS *Daley* (DD 519), 1952

# CHAPTER 2

## Triumph and Tragedy: The Space Race, Popular Culture, and Conflict

Though young officers writing New Year's deck log poems during the Cold War were vigilant toward the Communist threat, they never lost touch with the world around them. And the mid-20th-century world was changing fast. The optimism springing from the economic prosperity that followed World War II and continued into the early 1970s was tested by the social movements in the 1960s and brought to a halt in the recession of 1973-1975.

### Space

One of the most exciting arenas of change in the '50s, '60s, and '70s was outer space. Beginning with Russia's launch of Sputnik on October 4, 1957, the space race led to great scientific strides, culminating in the historic Apollo 11 moon landing on July 20, 1969. Though Sputnik was short-lived compared to later, more sophisticated satellites, its success as the first satellite launched from earth inspired the imaginations of scientists, musicians, and artists around the world. Young Officers of the Deck were not immune to "space fever," and their voices echoed the pride they felt in their country and in the Navy's critical involvement in their country's success.

Just three months after Sputnik's success, on the first watch of 1958 aboard USS *Yorktown* (CVS 10) at Peugeot Sound Naval Shipyard for overhaul, Lt. j.g. W. L. Fadner played with two possible meanings of the word "stirred" in his reference to Sputnik. Were the men in his log stirred from their sleep or stirred in their imaginations at Sputnik's launch?

> Units of the Reserve Fleet are sound asleep,
> But recently stirred at Sputnik's "beep beep."

On that same New Years night aboard USS *Antietam* (CVS 36), stationed at the opposite corner of the country, in Mayport, Florida, Lt. j.g. K. H. Wilson also mentions Sputnik.

> With the arrival of Autumn
> About the time Sputnik was born,
> Captain R. W. Cooper
> Relieved Captain W. A. Thorn.

Three days after Fadner and Wilson wrote their deck logs, on January 4, Sputnik would fall and burn in the earth's atmosphere after three months circling the globe.

The space race heated up in April of 1961 when the Soviet Union's Yuri Gagarin became the first human in outer space, circling the earth on one complete orbit. Three weeks after Gagarin's flight, the US rocketed astronaut Alan Shepherd into a fifteen-minute suborbital flight on May fifth. Gus Grissom followed in July with a second sub-orbital flight, and on November 29 of that year, the US sent a chimp named Enos around the earth on a two-orbit test flight.

Finally, on February 20, 1962, US astronaut John Glenn, a former fighter pilot who had flown 59 combat missions in World War II and 90 more in Korea, completed three orbits and became an American hero. Glenn would later serve as a US senator and in 1998, aboard the space shuttle Discovery at age 77, would become the oldest person to travel to outer space.

Ten months after Glenn's historic first orbital flight, on January 1, 1963, Lt. j.g. W. "G" Reid remembered being part of Operation High Jump, the recovery mission for Glenn's spacecraft. As OOD on USS *Brownson* (DD 868), Reid looked back at the experience. Glenn was cleared for each of the second and third orbits only as the previous orbit finished, and there was an electronic glitch with a warning light on the orbiter that created anxiety back at the space center for a dangerous reentry. Reid said,

> We waited and waited
> Glenn's orbit belated,
> Standing by in event
> The trajectory bent.

In the first deck log of 1966 aboard USS *Wright* (CG 2), Lt. Robert M. DeMais boasts of a significant accomplishment in the US space program.

> This was the year that brought about a rendezvous in space
> Between two capsules at top speed—the US won this race.

The rendezvous DeMais referred to was the first orbital rendezvous of capsules, which the US had conducted on December 15, when astronauts Frank Borman and James Lovell, flying Gemini 7, docked with Walter Schirra and Thomas Stafford on Gemini 6.

But the event generally thought to proclaim the US winner of the Cold War space race was the Apollo 11 mission, landing the first humans on the moon on July 20, 1969. When the Eagle landed in the Sea of Tranquility and Commander Neil Armstrong and lunar module pilot Buzz Aldrin stepped onto moon dust as command module pilot Michael Collins orbited above, the world was watching. Six months earlier, on December 21, 1968, the first great pre-moon-landing mission, Apollo 8, launched from Kennedy Space Center (the first to do so) with its crew of astronauts Frank Borman, Jim Lovell, and William Anders. It was the first human spaceflight to reach the moon, orbiting ten times without landing before returning to earth. Apollo 8 set the stage for the dress rehearsals—Apollo 9 in March 1969 and Apollo 10 in May—and for the actual moon landing with Apollo 11.

The success of the US in the space race was a source of hope and pride for all Americans during a time of significant upheaval in the country. The Vietnam War, the violence of the Civil Rights Movement, various other forms of social rebellion were straining the fabric of American society. But aboard USS *Denabola* (AF 56) at the beginning of 1969, J. E. Tankersley, Chief Storekeeper, seemed to express hope of things getting better, citing both the war and the space race.

> The world is in turmoil, on this the first day,
> Of a New Year that promises to speed quickly away.
> With astronauts planning to land on the moon
> And peace talks in Paris to end the war soon.

And on that same January 1, 1969, aboard USS *Sandoval* (APA 194), Lt. j.g. L. D. Newlon expressed the pride felt by the ship's crew after playing a central role in the recovery of the Appollo 8 capsule.

> In December we took part in a mission so great
> As a recovery unit for Apollo 8.

We're proud of the men who around the moon did zip,
And proud to have helped assure them a safer trip.

**The Poles**

American exploration of outer space was not the only mission sailors were proud of. During the Cold War, the US intensified its exploration of the Arctic poles. One of the important early operations in Antarctica was dubbed Deep Freeze by Admiral Richard Byrd, who had by the 1950s become the epitome of the exploration of the earth's poles. Byrd was the officer in charge of Operation Deep Freeze I in 1956-'57, which established permanent bases at McMurdo Sound, the Bay of Whales, and the South Pole. The flagship of that first Operation Deep Freeze was USS *Arneb* (AKA 56), whose OOD on January 1, 1956, was Ensign L. A. Higgins. The *Arneb*, which served in multiple Deep Freeze operations over more than a decade, was docked at Kainan Bay, where on January third the US would establish the fifth iteration of their Little America exploration base on the ice shelf. But on January first, Ensign Higgins mentioned that Admiral Byrd was at that time working from USS *Glacier* (AGB 4), an icebreaker that had opened the way to Kainan Bay and would then advance to McMurdo Sound to break ice for the establishment of the first permanent base there.

ADMIRAL Richard BYRD, Explorer,
On the GLACIER moored ahead
This expedition is directing
On 'Copter and power sled.

Ten years later, on the first watch of 1966 on that same ship USS *Glacier,* Ensign H. N. Harris toasts Admiral Byrd in the New Year's deck log from McMurdo Sound.

We keep our vigil for Uncle Sam,
And for Admiral Byrd who explored this land
Dear Admiral to you I propose a toast,
For on this ice shelf I will not roast.
'Cause in my veins the blood flows free,
'Tis for the United States and you and me.

**Civil Rights**

In addition to exploration of space and the arctic poles, officers of the deck on New Year's Eve mentioned many current events in American and international culture. The sailors listened to the same music and watched the same news broadcasts as their civilian peers, and the news of the day offered much fodder for their poetic imaginations. In the late 1950s the Civil Rights Movement in the US was heating up. The New Years deck log for 1958 on USS *Tweedy* (DE 532), written by Lt. j.g. W. R. Weatherhead, mentions the nuclear threat, the space race, and one of the signature events of the early Civil Rights Movement. In September 1957, nine young African Americans in Little Rock, Arkansas, who came to be known as the Little Rock Nine, attempted to desegregate Central High School and were met with violent protests by the public and prevented entry by the Arkansas National Guard. Only after President Dwight Eisenhower sent army troops to Little Rock were the students allowed to attend. It was a national story, and in the opening line of his deck log, written just three months after the crisis, Weatherhead was able to allude to it simply by using the name of the city.

Missiles, Sputnik, and a place called Little Rock

## Beatniks

Another social movement of the 1950s, known as the "Beat Movement," promoted spirituality and nonconformity and attacked the materialism of post-war America. Originating in the late 1940s, the movement developed in Bohemian communities in major cities around the country and had a significant influence on the decade of the 1950s and on the counterculture of the 1960s. Adherents were known as "Beatniks," and three of the best known and most influential writers of the movement were Allen Ginsberg (*Howl,* 1956), Jack Kerouac (*On the Road,* 1957), and William S. Burroughs (*Naked Lunch,* 1959). The Beats heavily influenced art and literature, creating a distinctive linguistic and social style. Their poetry was generally easily identifiable. In the first log of 1961 aboard USS *Walton* (DE 361), Ensign John D. Stempel's style is immediately recognizable as an imitation of the Beats.

> like, man, it's a cooool start for a hip ship, man—
> new year, new action, cat—the big six zero
> on the way out, cat, it was a real doped
> up time, cat . . . . .

The way Stempel uses words like "man," "cool," "hip," and "cat" make his parody work beautifully.

## Cyprus

In 1962 the island of Cyprus was in the world's eye, as tension between Greek and Turkish Cypriots created unrest, and the US worried about the stability and security of the region. On the first midwatch of 1963, Ensign H. D. Huffman, on the quarterdeck of USS *Biddle* (DDG 5), pondered the conflict.

It seems that Cypriots of Greek and Turkish mien,
Have differences to settle 'fore peace again is seen.
And we are standing by, alone, if e're the need arises,
To evacuate our countrymen to refuge from the crisis.

## Berlin

In January 1964, as Ensign G. D. Dieckmann wrote the New Year's log aboard USS *Koiner* (DER 331), his thoughts were on the events of the past year. At one point in the log, he focuses on a military desertion. Though Dieckmann does not refer to him by name, the subject seems to be Captain Alfred Svenson, who in 1963 became the first and only US officer to defect to the USSR during the Cold War years. On 4 May of that year, Svenson drove his jeep across the border into East Berlin and asked for asylum. In his log, Svenson describes the event as "an American boy who deserted his crew." A year later, in the first log of 1965 aboard USS *Hollister* (DD 788), a young Lieutenant, junior grade, manning the quarterdeck expressed his frustration at the continued division of the city of Berlin between East and West.

The Berlin Wall remains a disgrace—
Six feet of concrete to keep a city in place.

## Silver

A looming shortage of government stocks of silver in 1965, known as the Silver Crisis, led President Lyndon Johnson in July of that year to sign a bill that removed the silver from dimes and quarters and cut the amount of silver in half dollars. DeMais may be referring to the silver crisis in the following lines.

> Rising prices on metals almost made a mess
> But LBJ avoided that by selling his for less

**The Pill**

DeMais then turns his attention to an issue that had been causing a stir in the country. In 1963, Pope John XXIII established the Pontifical Commission on Birth Control in response to an unsettling development three years earlier that had created widespread debate in the church—the appearance of "the pill." The commission report would not be published until 1966, but on New Year's Day, 1965, Lt. DeMais wondered what the outcome would be.

> One point that Christians all 'round are contemplating still
> Is what the good Pope Paul will do about that strange new pill.

Pope John had been replaced at his death by Pope Paul VI. Though later in 1966 the commission would assert that artificial birth control was not intrinsically evil and that Catholic couples should be allowed to use it, Pope Paul would reject their findings and in 1968 issue the encyclical *Humanae Vitae*, reaffirming traditional teaching.

**Long Hair**

Observing the latest hairstyles for men, DeMais echoes a common reaction at the time, offering a light-hearted gibe at the "boys."

Music had a great year—the Barbers all got wealthy
With fancy hairdos for the boys
Which made them look unhealthy!

**Music**

He then turns his focus to the music. One of the longest-running American New Year's Eve radio broadcasts in history ran from 1959 to 1976—Guy Lombardo and his Royal Canadians hosted the show every year, which always closed with a rendition of "Auld Lang Syne." Lombardo, who would receive a citation from Congress for music that expressed hope, peace, and happiness for all mankind, would come to be known as "Mr. New Year's Eve." DeMais may well have listened to the broadcast before writing his deck log.

The strains of Guy Lombardo
We can almost hear
As they tell us of the coming of a grand and hopeful year—

**Sports**

On the same night that DeMais offered his cultural summary, Lt. j.g. Michael Fanin, the OOD aboard USS *Hammerberg* (DE 1015), revealed that he was a sports fan. After complimenting Secretary of Defense Robert McNamara, DeMais mentions a famous Soviet shot putter and discus thrower who won multiple gold medals in the 1960 and 1964 Olympics and two famous American football players—

. . . . Strike up the band
For our leader, SecDef Macnamara

And our favorite shotputter, Tamara
Press. Revelers bend your elbows for Babe Parilli,
A poor guy too often knocked silly
By Ernie Ladd . . . .

Babe Parilli, a two-time All-American quarterback at Kentucky and eventual Hall of Fame selection, led the Boston Patriots of the American Football League for most of the 1960s. Ernie Ladd, a defensive tackle for the San Diego Chargers of the American Football league during the 1960s, could very well have knocked Babe Parilli silly during that time. His nickname was "the Big Cat." Ladd, who was 6' 9" and 290 pounds performed as a professional wrestler in the AFL off-season and went full time as a wrestler when he retired from football. He was inducted into the Chargers Hall of Fame in 1981 and was inducted into four different world wrestling halls of fame (WCW, WWF, *Wrestling Observer Newsletter*, and NWA) from 1994 to 2013.

**1968**

In the first log aboard USS *Bexar* (APA 237) in 1969, Lt. j.g. L. C. Brown, in poetic prose, looks back on one of the most disruptive years in American history, with powerful events happening not only in the US but around the world. The month of January 1968 alone included the "Prague Spring," North Korea's capture of USS *Pueblo,* and the Tet Offensive in Vietnam. But the next four months saw the assassination of Martin Luther King, Jr. in Memphis resulting in riots across the country, the first of several gun battles during the year between police and members of the Black Panther Party, the massive rise of student protests in the US and around the world leading to the May 6 "Bloody Monday" in Paris, and the assassination of Robert F. Kennedy in June. August saw the crushing of the Prague Spring by

Soviet troops and the violence of the Democratic National Convention in Chicago, followed in October by the Black Power demonstrations by US Olympic athletes during the National Anthem. Lt. j.g. Brown alludes to many of these divisive events in his short summary. He also notes, on the positive side, the Apollo 8 mission, the beginning of peace talks, successful heart transplants, and the unity surrounding the Olympics as a whole.

> Three hundred and sixty-five days ago a new year began. It was a year of external and internal conflict. Our longest war continued another year. There was bloodshed in the cities. Student riots on campuses. Leaders slain in the streets. But there were achievements also. Three men orbited the moon. Peace talks were started. Prisoners were released. Hearts were transplanted. And for a few days men forgot their differences and competed in athletic events as equals, sharing in their defeats and their achievements as brothers.

Brown ends the log on a hopeful note by quoting Thomas Paine:

> The world is my country
> All mankind are my brethren,
> To do good is my religion.

Also in January 1969, the log of USS *Bluegill* (AGSS 242) rings a hopeful note. In it, Lt. J. R. Sadler, expresses gratitude for the return of the *Pueblo* crew on December 23 after eleven months of captivity, torture, and interrogation in North Korea.

> Our long awaited Pueblo's
> Officers and crew
> Are finally home.
> Boy just what have they been through!

**Recession**

In February and March of 1973, 591 prisoners of war were returned to the US from Vietnam, marking the end of a long emotional struggle for their families and the country. In October, a shortage of oil caused by an embargo by Arab oil-producing states aimed at countries that had supported Israel in the Yom Kippur war earlier that year created an international crisis. In the US, gas shortages and long lines led to President Nixon recommending in November a national speed limit of 50 m.p.h. for passenger cars and 55 for trucks and buses. The restrictions were eventually enacted for all vehicles at 55 m.p.h. The recession caused by the oil embargo stretched from late 1973 into early 1975. Aboard USS *Alamo* (LSD 33) on the first watch of 1974, Command Senior Chief K. C. Banta ponders some of these struggles but also offers a selection of positive achievements.

> The speed limits have been lessened more folks ride the bus
> Many luxury's we took for granted, are no longer available to us
> 1973 is now fading from sight
> As we look back now many things happened right
> There was Skylab 1 and Skylab 2
> In 73 our armed forces honorably withdrew
> The nation stood quiet as our POW's came home
> In their hearts they knew they were never alone.

**Pop Culture: Hula Hoops**

Throughout the ups and downs of the Cold War years in America, the deck log poets, as has always been the case, never lost their sense of humor, and they stayed in close touch with current popular culture. One of the most interesting and enduring fads of the 1950s was the Hula Hoop, popularized in 1958 by the Wham-O toy company, which sold 25 million hoops in the two months after release.[1] On the opening watch of 1959 aboard USS *Rowe* (DD 564), Lt. j.g. R. D. Mitchell mentions "hoola" hoops and the satellite craze.

> Yes, old shipmate, looks like another year has flown by, / One that sported hoola hoops and things in the sky.

**Elvis**

Mitchell goes on to refer to another huge cultural sensation of the 1950s. Elvis Presley's career started at Sun Records in 1954, and it was soaring when, in March of 1958, he was drafted into Army. It was a massive media event, with television cameras covering everything from his physical exam to his first buzz cut. The continued attention to Presley's status in the Army shows up in Mitchell's comic line about him.

> And even Elvis went where all good doggies go. / By and large I'd say that things weren't any too slow.

**The Kingston Trio**

In 1959, folk music rose to popularity. The Kingston Trio recorded the album titled *from the Hungry i*, which was awarded a RIAA gold album in October of 1960. One song in particular, though comic, included a catalogue of the current events that fed the angst of the Cold War. Its official title was "The Merry Minuet," written in 1955 by Sheldon Harnick, but after the Kingston Trio recorded it, it came to be known more popularly as "Rioting in Africa," from the opening line of the song. Two and a half months after the Kingston Trio were awarded the gold album in October of 1960, Ensign D. R. Atkinson welcomed 1961in his log aboard USS *Lenawee* (APA 195) with a reference to the song.

> This year opens with things a little more complicated than they were several years ago when according to song they were rioting in Africa.

**Mission Impossible**

From 1966 until 1973, a popular television series titled *Mission Impossible* reflected the real-world covert struggle of spies and secret agents in the Cold War. One of the most memorable phrases used in each episode, a phrase that has outlived the show and can still be heard among a generation of watchers, was "This tape will self-destruct in five seconds." The lead character, the American spy Jim Phelps, was given his assignments on a small tape recorder on which the tape self-destructed after the message was received. On January 1st, 1969, in the middle of the eight-year run of the show, Lt. j.g. Richard W. Deming, OOD on USS *Woodpecker* (MSC 209) 1969, ends his log with the iconic line.

> This log will self-destruct in five seconds. Good luck, Mr. Phelps.

# SIDELIGHT 3

## Should old acquaintance be forgot...: A Midwatch Poet Responds

Several years ago, one of the authors (Dave Johnson) began locating family members and/or the writers of midwatch poems. He sent emails, letters, or social media messages containing the midwatch poems to these unsuspecting recipients. Their responses were overwhelmingly positive to seeing their family member's poem or a poem that they had written so many years ago.

Ken W. Sayers[1] wrote a response about receiving a copy of the poem he wrote for the USS *Walton's* (DE 361) New Year's Day midwatch log in 1966. Here is his response:

> There I was, on November 1, 2025, going about my usual daily business when I received a letter that was out of the blue and far from routine. The letter informed me that a document I had written on January 1, 1966, was going to be included in a forthcoming book. Wow, what a surprise! And attached to the letter was a photocopy of that document, which I immediately recognized from its words and the style of its printed letters. As you know from the book

you are now reading, the document in question was an entry that I had penned 60 years ago in the deck log of a U.S. Navy vessel, and, in accordance with naval tradition, because of the special date and time, the entry was in the form of a poem. Not a rhyming poem, in my case; instead, it was rendered in free verse.

As I read the document I was, of course, immediately taken back in time to when I was just 23 years old and serving as new ensign on a ship that had been commissioned two years after I was born. So naturally I thought then and again now about my shipmates. We had a small crew—just three officers and 40 enlisted personnel—and I got to know all of them very well. I learned a great deal from those men, especially the senior petty officers who had considerable experience at sea and in their individual ratings. They were experts about their equipment and operations, while I was still finding my sea legs and beginning to understand the basics of my responsibilities. If you're smart as a newly commissioned line officer, you spend a lot more time listening to your chiefs and senior POs than in giving orders.

During the two years I spent on the ship I also learned about other parts of the country from the men who came from all over the United States, and who together constituted a unique and diverse community of our own. We had different origins, personalities, experiences, attitudes and accents. But fate and the Bureau of Naval Personnel brought us together, and each of us worked together successfully while fulfilling our respective roles. I guess what I was trying to accomplish in that midnight log entry so long ago was to [a] record the information required by Navy Regulations regarding the ship's status at that time and place; and [b] acknowledge and provide the context of the crew's contribution to the nation.

As I was reminded by Dave Johnson's letter, and as I look back on it now, the time I spent on USS *Walton* was among the best years of my life. When someone learns that I am a Navy veteran, they will often say "Thank you for your service." While appreciated, those words also make me feel uncomfortable because in my head I'm thinking: "No, I need no thanks. I'm just thankful that I had the opportunity and privilege to serve as I did." Moreover, I am reminded again that I miss those days on the ship. I have often wished that I could step back on her decks once again, even for just one day. But I can't and never will. That's because *Walton* was expended as a target on August 7, 1969, and sunk off the coast of southern California. Thus, I am today thankful for the experiences and the memories I have, and grateful to the authors of this book for preserving some of them in print.

Here is Ken Sayers poem:

Dark of the darkest dark,
Water of brine content,
Currents tug and shove
And pull on old Walton.
The deepest joy of a new-born year,
And I, exhilarated at being alive,
Shiver in January's nativity and pace
My quarterdeck and
Record the instances and the facts
Surrounding life as it breathes aboard
This vessel, this day.

An island in San Francisco Bay,
A flat and artificial and unreal island
Treasure Island it is called and an
Extension of that collection of soil
Pier 14 runs its skinny finger up along our
Long grey hull, tied to it with six standard lines
And it opens its pores and it
Receives invisible scalding steam and
Water.
Condition Yoke Modified is set.
Down beneath metal decks and sheltering bulkheads,
Laying on canvas supported by chains
Dream, rest, sleep, toss, exist,
Forty White Yellow Black men.
Red lights glow along empty passageways
Guiding this watch — watch condition V — as it walks about the decks
And out over Walton's rail and sitting at an adjoining pier
Graceful, trim and cat-like is Twining
Close too, is the obese transporter of troops, Blatchford
Here and there tug boats motionless, silent, Indian names.
Beyond is darker than the night's own dark is the Bay Bridge.
Cars on their way to celebration or running away from noise and party
Move soundlessly across the looming structure.
Still beyond even that a revolving white light shows Alameda,
SOPA they call him, is Commander Western Sea Frontier.
Why sit here on this black and unrewarding morning?
It is cold and the twinkling icicles of stars above lights

On the shore inquire as to why.
And why is a question deserving of a real and meaningful
Reply
In warm homes this clear crisp New Year's sleep the millions
Who call themselves "protected" and yet citizens must protect
Their own and relinquish comfort for this safety we need.
And there are some who have done this scene but seek to
Replay it at every opportunity and they, citizen-soldiers,
Sail in Walton monthly and prepare for a deadly re-run.
The moon and the red truck lights play counterpoint to the
Steady radiation from the pier lights and Coke machine.
Looking at all of this, feeling the small gentle roll,
Smelling the oil and the soap and the salt air,
I am glad.

# CHAPTER 3

## Wine, Women, and . . . .

So here I stand on New Year's Eve,
No wine, no women, no song.
An unwilling martyr for all to see,
Oh, this Navy has done me wrong.

--Ensign David R. Carlson

USS Sabine (AO 25)

January 1, 1961

Wine, women, and song have long been the province of poets, composers, and fiction writers. They have appeared together in doggerel, serious poetry, and tunes; in English translations of German sayings; and even in medieval Latin verse.[1] And, as we have seen, young U.S. midwatch poets occasionally intended their logs to be sung to popular melodies. Many more used the midwatch log to extoll the presence, or (more often) lament the absence, of alcohol and women—usually girlfriends or wives, but sometimes women in general.

## Here's to the New Year

Alcohol and members of the military have a long, intertwined history. As early as the 1750s, colonial soldiers received a gill (about a half cup) of rum per day; and when a chaplain complained that the men were not faithful in attending prayers, it was Benjamin Franklin[2] who proposed solving the problem by serving the rum only after prayers.

By 1794, the U.S. Congress had officially established a daily alcohol ration for sailors aboard Navy ships. The ration was set at a half-pint of distilled spirits, or a quart of beer.[3] Then, on July 1, 1797 the ration changed to simply one-half pint of distilled spirits, likely due to the fact that beer did not preserve well, nor maintain its taste as well as higher-proof liquor.[4] Rum remained the standard issue until 1806, when the Navy began encouraging whiskey as a cheaper alternative, and in 1842 the U.S. Congress reduced the daily ration to one gill and decreed that no sailor under the age of twenty-one should receive the ration.[5]

Throughout the remainder of the nineteenth century, Navy rules concerning alcohol changed from time to time until July 1, 1914, when General Order No. 99 prohibited alcoholic liquors aboard ships and in Navy yards and stations. Following the 1933 ratification of the twenty-first amendment repealing prohibition, the Navy polled flag officers in an effort to decide whether the alcohol ban should be rescinded; their conclusion was that it should stand.[6] One result, no doubt unintended, was that the absence of alcohol became a recurring theme in New Year midwatch logs.

Ensign D. T. Leugers cleverly illuminated the dry sailors' dilemma as the Year of the Rabbit (1963) gave way to the Year of the Dragon (1964) aboard USS *Castor* (AKS 1). Ensign Leugers wrote in good humor and with an oblique reference to the Greek poet Homer:

The rabbit has become a dragon
So down with work and hoist a flagon,
But woe to him who has the watch
He'll have to do without his scotch.

Enough of this! It's time to write
The log in verse and seem real bright.
To write in verse is not too hard
If you happen to be a blind Greek bard.

**Water, Water Everywhere**

When his ship is at sea, the OOD's lament often encompasses not just those with the watch duty, but the dry condition of the entire ship. Thus, as 1973 dawned, and the ammunition ship USS *Butte* (AE 27) steamed around Cuba, en route from Norfolk to Panama, Ensign L. P, Belote wrote:

Course 159° T speed 20 kts, SOPA is on board,
No liquor allowed for the good ship's horde,
Material condition YOKE, readiness condition IV have been set.
By no booze this eve will our taste buds be wet.

Earlier, as 1956 arrived on a night that was freezing cold, Ensign J. H. Dole surely expressed the sentiments of his shipbound mates on USS *Remey* (DD 688):

On this frigid, frosty night with the temperature twenty-seven,
On the quarterdeck of the REMEY, that old destroyerman's heaven,
We are conning in the New Year, shoving off the old—
But we'd feel a bit more chipper with some brew to nip the cold.

In the same vein, Lt. R. J. Casey bemoaned the fate of his crew as the destroyer USS *John R. Craig* (DD 885) cruised toward the Philippines while 1964 arrived in the Pacific:

> From Hong Kong we came, to Subic we go. None can
> Deny it's a shame,
> But good things will come and good things must go and
> No one is really to blame.
>
> Condition YOKE and readiness FOUR is set throughout
> The whole ship.
> The plant is split on one and three, and I'm 'bout ready for a nip,
>
> For a brand new year is barely here but dry we
> Certainly be.
> A dismal note to be afloat on the night of
> New Years eve.

Sometimes the midwatch log is not light-hearted, as in this entry by Lt. j.g. B. C. Brooks as USS *Holder* (DDE 819) cruised near Cuba:

> New Year's day here is quieter than most,
> Since the crew must diligently remain at their posts
> while steaming east 'long Cuba's Coast;
> Still coffee's raised high in a 'sober' toast.

On the other side of the continent, on January 1, 1958, the submarine USS *Caiman* (SS 323) was bound for Japan when Lt. R. M. Smith logged this report (safety and negative are tanks filled with water to reduce buoyancy, allowing the sub to sink deeper; *akadama* is a sweet red Japanese wine):

Safety and negative are flooded; we're ready to submerge;
All we need is the order when the Skipper gets the urge.
Chief Lissoon has the watch well in hand down below
With twenty stalwart men to make the submarine go.

Oh for a 'scotch' or a can of cold beer
Instead of black coffee to bring in the New Year;
But we'll patiently wait till we get to Yokohama
For an overdue 'hot bath,' and much 'akadama'

A year later, Caiman was again bound for Japan, from San Diego via Pearl Harbor, and Lt. Smith once again drew the midwatch duty, this time mentioning the Golden Dragon, a reference to the unofficial award given crew members who cross the International Date Line:

From San Diego we've been torne,
To Hawaiian shores we're borne,
And thence to the realm of the Golden Dragon
Where Nippon beer will fill our flagons.

On January 1, 1961, USS *Tracer* (AGR 15), a radar picket ship reconfigured from a Liberty Ship, was bound for San Francisco. OOD Lt. j.g. M. P. Spillum perhaps summed up the sentiments of all those at sea on the first holiday of the new year:

And when we are back on the beach once more
With our shipmates, loved ones and friends
We'll fill up our glasses with only the best
And properly welcome a new year again.

## It's a Fine Place for Sailormen

At the turn of the twentieth century, poet John Masefield wrote, "It's a fine place for sailormen is that there port. 'N' I wish—I wish I was there." Many a sailor has wished to see his home port, or any port, especially on such a celebratory occasion as New Year's Day. Our midwatch poets were no exception, but, having drawn the OOD duty, they had to watch and wait as their compatriots went ashore to ring in the new year. Their feelings about this have often made their way into midwatch verse, as we see in the opening of this log, penned by Lt. j.g. R. H. Boyd of destroyer USS *Picking* (DD 685) while moored at Long Beach:

> Here we are on New Years Day
> Silently passing the time away,
> Thinking of our friends out on the beach
> With liquid spirits in easy reach.

Ensign M. J. Schneider had similar feelings while his shipmates from the dock landing ship USS *Plymouth Rock* (LSD 29) brought in the new year ashore in San Juan, Puerto Rico:

> I'm staying aboard on this New Years Eve
> My shipmates are celebrating but I can't leave.
> I can't bring in the year with a snort or a nip
> 'Cause black coffee is all we're allowed on the ship.

Coffee was also the lone drink aboard the minesweeper USS *Firm* (MSO 444) as Lt. j.g. B. L. Douglas welcomed 1968:

> In town, wine is flowing like hot molten lava
> But here we'll be toasting with a cup of java.

Lt. J. Scott Kirkwood managed to invoke some humor as he imagined the onshore antics of the crew of USS *Karin* (AF 33), living it up at Pearl Harbor while he stood the midwatch:

The crew's gone ashore, and I am thinking
That by now they're probably all stinking.
Not to infer that this group does much drinking!
They just enjoy the sound of glasses clinking . . .
A word of advice for all of you guys
Who'll return with big heads and bloodshot eyes:
Drive safely, be careful, use common sense!
Ninety percent of the people are caused by accidents!

In a concise conclusion to his 1964 midwatch log, Electrician's Mate R. W. Headrick of attack cargo ship USS *Bellatrix* (AF 62) expressed the regret of many an OOD as he greeted the new year:

No hangover tomorrow
Very much to our sorrow
We started the New Year today

And Lt. j.g. G. R. Hallberg of USS *Boston* (CAG 1) ushered in the new year with a rather unusual rhyme scheme in a self-effacing lament that ended like this:

And yet there is hope as I write this new page,
Struggling to earn my quite minimal wage
Tomorrow's a new day if not a new year
And Boston's a lady who never shows her age.

And that, my good shipmates, is all of the news.
So if in your heart there's no place for this muse
Pause to consider that I'm a short-timer,
Who's never pretended to be a good rhymer,
And that I would rather be quaffing the brew
Than having to write this poetical primer."

**Splicing the Mainbrace**

In the days of true sailing ships, huge sails were attached

to poles known as yards and maneuvered by ropes known as braces. The longest and most important of these, the mainbrace, could be as large as twenty inches in diameter, was essential to maneuverability, and was thus often the target of enemy gunners. If the mainbrace was damaged in battle, repairing it required both precise skill and strength; effecting repairs in the heat of battle was no mean feat. Due to the arduous nature of this task, and its importance, it became customary, when the job was complete, to give the crew an extra ration of rum.[7] The order "Splice the mainbrace," dating from at least 1773, eventually came to mean simply having a drink, especially following any harrowing task, or on special occasions.

British poet John Masefield, having been to sea himself, captured the essence of the tradition, following a near tragedy, in 1913 in a stanza from his epic narrative poem *Dauber*:

> The watch was called; Dauber might go below.
> 'Splice the main brace!' the Mate called. All laid aft
> To get a gulp of momentary glow
> As some reward for having saved the craft.
> The steward ladled mugs, from which each quaff'd
> Whisky, with water, Sugar, and lime-- juice hot,
> A quarter of a pint each made the tot.

This old tradition was not lost on our midwatch poets, who bemoaned the lack of opportunity to splice the mainbrace themselves. Thus, as the year 1957 arrived aboard USS *Newport News* (CA 148), Lt. j.g. R. B. Lawson found himself envying those who did not have the misfortune to stand the midnight watch:

> So this log I will close and relinquish my place
> As I enviously watch those who have spliced the mainbrace.

Lt. j.g. Harold E. Paxton found himself in similar straits aboard a very quiet USS *Corsair* (SS 435):

North of pier seven,
A dead quiet place,
For most are out
Splicin' the 'brace.

Lt. P. J. Vogelberger, Jr. may have been even more morose as he stood the watch on USS *Snook* (SSN 592):

The duty section its vital watches stands
With heavy heart and disappointed face,
Regretting the lost chance to join all hands
In a New Years soiree to splice-the-mainbrace.

And, moored at Chicago, Lt. H. R. Moore welcomed the new year to USS *Daniel A. Joy* (DE 585) and suggested splicing the mainbrace with an alternative to alcohol:

On the streets of the city a celebration in progress,
But duty bound, the watch must not digress.
This New Year's on beverages alcoholic we may not choke,
So will "Splice the 'Main Brace'"- Desig. COKE.

Not every midwatch poet found cause to grumble about drawing the midnight duty and missing the opportunity to splice the mainbrace. As a very dry USS *Bennington* (CVA 20) steamed from Yokohama to Pearl Harbor, Lt. J. E. Liles looked forward with good cheer as he extended wishes for the coming year:

At zero one hundred as we write this rhyme
The clocks are advanced to zone Kilo time.
The gasoline system's reported to be
In normal condition at quarter past three.
The waves roll us gently, the wind sings so free

Of the pleasure awaiting us home from the sea.
So splice ye the mainbrace all men of good cheer
The BENNINGTON sends you a HAPPY NEW YEAR

## My Girl Back Home

In *South Pacific*, the famous Rodgers and Hammerstein musical, Lt. Joseph Cable and Navy nurse Nellie Forbush sing "My Girl Back Home," while missing faraway loved ones in Arkansas and Pennsylvania. In the Navy, women served as nurses as early as Civil War days, and became an official part of the Navy when the Navy Nurse Corps was established in 1908.[8] However, our midwatch poets write, not about nurses, but about the wives and girlfriends they are missing on New Year's Eve, and the good times their shipmates are having ashore with female companions.

The old saw about sailors having a girl in every port is, of course, a myth, but the trials of lengthy separations occasioned by extended deployments are real, and have existed for as long as sailors have gone to sea.[9] In the eighteenth and nineteenth centuries, some sailors remembered their wives and girlfriends with sailors' valentines. These "valentines" were actually small, hand-crafted boxes—sentimental souvenirs given by sailors to wives and other women; despite their name, they had no real link to St. Valentine's Day. Many were made not by sailors themselves, but by craftspeople in Barbados, the last port of call for many homeward bound seamen of the era.[10] And sailors have also, of course, written not only log poems, but love letters to those far away at home.[11]

**Girls, Any Girls!**

Although sailors' valentines were a thing of the past, U.S. Navy officers standing the midwatch during the Cold War years nevertheless missed their wives and girls just as did their earlier counterparts. Sometimes, however, their laments were more frivolous, and not necessarily directed to wives and girlfriends, as in this midwatch log penned by Ensign F. R. Stein on USS *Banner* (AKL 25):

> The ship is in condition of Readiness V,
> (On the beach, the wild parties are really coming alive.)
> With PACOM DEFCON IV set,
> The babes in a frenzy waiting to be met.
>
> With material condition YOKE modified set throughout the ship,
> (It is useless to stand here, a military gyp.)
> If I were ashore, the girls would flip.

While USS *Harwood* (DDE 861) was moored at Miami, Lt. j.g. J. M. Taborn wrote about the girl who seems to have got away:

> Other ships present are 'merchies' and yachts,
> Where the girls are offering "forget-me-nots".
> The ship is secure, Yoke (modified) is set
> and a shipmate is squiring that gal I just met.

And there were, of course, no girls to be seen on a submarine engaged in special operations, as Lt. William A. Owens noted on USS *Seadragon* (SSN 584):

> Tis New Years eve on special ops.
> No girls tonight or kiss of the hops.
> While our mates ashore toast the New Year hour,
> We steam full speed on reactor power.

As we might expect on a lonely midwatch, the OOD's mind may wander, as Ensign C. P. Driscoll's did as 1958 arrived on USS *St. Clair County* (LST 1986):

> What a shame!
> I'm not to blame,
> For New Years duty
> Without a dame . . .
>
> This festive night
> I spend dreaming,
> Of femmes far off,
> And moonlight beaming.

On New Year's Day 1952, USS *Deuel* (APA 160) was moored at Norfolk Army Base, far from the war in Korea. Nevertheless, OOD Ensign D. A. Durfee found reason to wish for better company as he greeted the new year:

> Our company is not a maiden, loving, gay and sweet
> But rather various units of our gray Atlantic Fleet.
> Another year has thus begun, so cheer up, don't be blue'
> The Deuel again shall see the world in 1952.

Ten years later, female company was also absent from USS *Acme* (MSO 508), as reported by Lt. j.g. G. E. Jones:

> With the stroke of twelve the old year passes,
> We'll welcome the new without our lasses.

And at Orange, Texas, less than two months before his ship was decommissioned, Lt. j.g. G. Cohen of USS *Kenneth M. Willett* (DE 354) mused about the whereabouts of local girls:

> At U.S. Naval Station, Orange, Texas,
> The watch is wondering where the other sex is.

The midwatch can be lonely, as Lt. j.g. A. L. Gottschchalk, Jr., observed on the submarine USS *Thornback* (SS 418):

> T'was raining the evening that '61 died.
> All watches were stood with Navy pride.
>
> Deep down in our hearts we were all willing;
> to see that the New Year received top billing.
>
> Most of our shipmates were having a ball;
> while down on the boat we could have none at all.
>
> Not even the cheer of a feminine kiss
> Just reminiscing in silent bliss.
>
> The shipyard in Charleston was quiet and lone
> while we sat in dry-dock dry as a bone.

**The Girls They Left Behind**

While many midwatch logs reflected the general absence of women and the opportunity to celebrate the new year with them, others spoke of the wives and sweethearts who were on the minds of the midwatch poets. One of those was Ensign N. E. Harper, who was thinking of his wife as he wrote this 1962 log aboard USS *Aludra* (AF 55):

> It's New Year's Eve, and here I stand;
> What a heck of a place to be,
> Wishing that I were on dry land
> Partying—just my wife and me.
>
> 'Tis quiet, like every other mid,
> But the difference to me is great.
> If only this watch I could have rid,
> I would be in a happy state.

Similar sentiments prevailed on USS *Waccama* (AO 109), as Ensign J. V. Huyck wrote his midwatch log:

COMDESANT IS SENIOR HERE
AND USUALLY STAYS ACROSS THE PIER
BUT PROBABLY HE'S AWAY TONIGHT
WELCOMING THE NEW YEAR RIGHT.
MY WIFE AND I WOULD DO JUST THAT
IF I WERE HOME—BUT HERE I'M AT—
AND SO I SAY, ALTHOUGH I GRIEVE,
'HAPPY NEXT YEAR'S NEW YEAR'S EVE !!'

As USS *Douglas A. Munro* (DE 422) steamed toward Hong Kong, Lt. j.g. J. C. Froid looked ahead, and wrote of the loved ones awaiting at Pearl Harbor and the stories the sailors might tell them:

On the morning of the third with all our colors flying
We will be in Hong Kong Harbor lying

From there we steam to the land of the Rising Sun
To see if we can't all have a little more fun

For after Nippon we head back to Pearl
To settle down with wife or with the steady girl

To whom we will try, with some misgiving
To prove we have been doing nothing but good clean living

Sometimes a midwatch poet has worried about being stuck with the duty while another sailor was out with his best girl—a dilemma summarized by Lt. j.g. D. D. Niblo, while serving as OOD aboard USS *Caliente* (AO 53):

Its hard to compose a work of beauty
When its New Years Eve and you've got the duty.
Some other guy is out with your gal,
Could be a shipmate? A real swell pal!

> They're living it up at parties with booze,
> I'd better stop now or I'll blow a fuze.

For Ensign J. W. Lester and the men of USS *Dunn County* (LST 742), as 1956 dawned, a photograph had to do:

> A drink from the scuttlebutt will be our toast;
> Kiss the picture of our best girl.
> What more could a rugged sailor want
> To start the year with a whirl.

And four years later, as USS *Plumas County* (LST 1083) lay at anchor at Long Beach, Ensign J. A. Sand, III wrote wistfully:

> For instead of your arms and the smile I hold dear,
> The sight I behold surrounds the Twentieth Pier.
> The cold and the wind were as knives in my back,
> O, how could I be a man of good cheer? . . .
>
> . . . With MINPAC near by, 'twas not that I knew
> That the night would end all, before dawn would ensue.
> I lay there secure, with my hatches all tight,
> And dreamed of the New Year, and the nights I'd spend with you.

Despite missing loved ones, midwatch OODs nearly always found a way, in the end, to convey good wishes for the coming New Year, as in this log by Lt. j.g. R. P. Means, writing from the deck of USS *Aspro* (AGSS 309):

> The crew with the duty, the married, the single,
> Tho not with their families, they join in the jingle.
> From the ASPRO to all, a very pleasant new year.
> May God give us peace and live not in fear.

## And to All a Goodnight

Although they may well have been missing wives, girlfriends, and a celebratory toast, the midwatch poets nearly always found a good-humored way to mark the departure of the old year and the arrival of the new. They managed this despite cold nights, far-flung seas and ports, and the knowledge that shipmates ashore were having a fine time—or with hope that their ship, if at sea, would soon find its way to a port.

As he said good-bye to 1956 from the deck of USS *Remey* (DD 688), Ensign J. S. Hellewell struck an international chord:

> So here's to France
> where the ships of the line
> bring the New Year in
> with a glass of wine.
>
> And here's to the British
> out at sea
> enjoying a fortified
> brand of tea.
>
> And here's to Americans
> standing watch
> and damning the rule
> that denies them scotch . . .

And Lt. j.g. C. A. Heimbold, after wishing his USS *Gainard* (DD706) shipmates a happy new year, concluded his log in this way:

> So wish us well
> That next year we
> May pledge your health in more fitting glasses
> While surrounded by admiring lasses.

Lt. j.g. H. A. Ballenger struck a poignant note in his poem ushering in 1959 aboard USS *Brister* (DER 327):

> Here we are in cold and rain,
> how can we endure such pain.
> Condition Yoke or Baker set.
> Desired condition? You can bet,
> We wish that we were somewhere else,
> mixing drinks with Scotch and seltz.
> Although we are far at sea,
> we wish that we were home and free.
> To live it up with joy and mirth,
> But all I have is an upper berth.
> Happy New Year to the CNO,
> New Years here is mighty slow.

In the end, perhaps Ensign F. J. Mc Gimpsey spoke for all his fellow midwatch OODs as he watched 1958 roll in on a cold night aboard USS *Rhodes* (DER 284):

> Flank speed to Argentia and even quicker,
> This New Year's Eve without any liquor.
> The parties we all attended, the women we kissed.
> Are now fond memories—sorely missed.
> The old salts remember last New Year's Eve,
> The RHODES rocked and rolled; she sure did heave.
> At that time station Fourteen was our post,
> 'The roughest picket to date,' the old salts boast.
> Message 310325Z from Commodore LAIRD,
> Said in Argentia, your radar will soon be repaired.
> Zero-Six-Seven is the course we are steering,
> The cold northern waters we forever are nearing.
> Time 0024 changed course to Zero-Four-Nine,
> All ahead flank with all four on the line.
> We have set condition of Readiness Four;
> and material condition Baker once more.
> Recently we passed old Sable Island Light,

which confirmed our position on this dark night.
Security reports that throughout the ship,
all is well for another long trip.
Tonight the seas are calm, the temperature's cold,
This midwatch standing can make a man old.

# SIDELIGHT 4

## Ships as Women: Personal Pronouns She/Her

For centuries, maritime tradition gendered ships by referring to them as "she." We do not know exactly when this tradition began or why. A variety of suggestions emerged over the years, many of which are not necessarily mutually exclusive. Some point to the fact that "ship" in Latin is "Navis," a feminine-gendered noun. Others suggest that because mariners were primarily male, they viewed their ship as female as a nod to the vessel's nurturing character during their often-dangerous voyages on the high seas. Their ship kept them safe just as their mothers did when they were young and vulnerable. Elaborately carved figureheads adorning sailing vessels often portrayed beautiful women or mermaids possibly in the hope that their beauty would appease the gods of the seas and calm troubled waters.[1, 2]

It should not be surprising that in more modern times ships continue to be feminine-gendered and that this tradition is common in the content of midwatch deck log poems. While many poets offhandedly refer to their ship as *she* or *her*, some take their gendered personifications further than mere pronouns.

In 1962 aboard USS *Henrico* (APA 45), Lt. j.g. J. Meyer's log is written from the first-person perspective of the ship that is unequivocally female.

> I recall traversing the sea's wide reaches; the Normandy and Okinawan beaches; and that innocent young cargo that became men; trained, by grimness, the truths that war teaches, their blood, straining sand of foreign beaches; from mothers' breasts, never to be children again; I watched it all in those years, horrified; aged, I calmly set, at YOKE (modified).
>
> I senilely receive services from the pier, dreaming hopefully of a true new year's day, when, with peace of mind, and smile, one can say; there is, at last, among all men, no war this year; a tired lady needs her rest: she has passed her history test.

The Navy employed a variety of ships as **tenders** for other vessels (e.g., destroyers and submarines). These ships provided logistical support for other types of vessels. A ship caring for the needs of another easily conjures images of the traditional nurturing role of mothers. Chief Boatswain's Mate A. F. Cook aboard submarine tender USS *Fulton* (AS 11) in 1966 directly evokes imagery of this role.

> While Bang at our bow, (SS three Eight Five) now
> Has Entemedor secured with lines short.
> Usually more of the boats are snuggled like shoats
> Around Fulton, the mother so tender.

Chief Cook's use of "shoats," or newly weaned pigs was a nod to the term "pig boats" used as a derogatory name for submarines during the early 20th century.

Even the harbor was viewed as an entity that possessed the characteristics of women as Lt. j.g. S. B. Haug, USS *Gregory* (DD 802), 1962, wrote.

> The pier the placenta, the harbor our womb.
> Patient and calm we wait out the days
> 'til we are born to the sea our powers all abloom,
> To defend the peace in our rigorous ways.

Some OODs elected to go beyond the obvious imagery evoked by ships' maternal nature. Lt. C. N. Osborne aboard USS *Fulton* (AS 11) moored in Boston in 1958 awaiting significant repairs wrote:

> The Fulton floats a disheveled queen
> Here in the home of cod and bean
> Pier two is where she has her berth
> Waiting the day to reprove her worth
> Moored portside to, her lines are doubled
> Though outwardly calm, she's inwardly troubled.
>
> ..........
>
> Her paint is patched, her insides torn
> She's looking sad and feeling forlorn
>
> ..........
>
> But soon the yard work will be o'er
> And then the FULTON will leave this shore
> To her will fall new tasks a myriad
> For the Bureau will know she's had her period.

Following in a similar vein we find a log entry acknowledging another aging ship.

> Out of the frying pan and into the fire
> Is the ships own power as our generators tire
> Alas the Cheboygan is a middle-aged lady

Who suffers many a menopausal malady.

J. M. Stemmons, Lt. j.g., USNR

USS *Cheboygan* (LST 533), 1969

Edwin C Link, Lt. j.g., described in detail the "surgeries" required by another ship who is getting on in years. The poem is from the log of USS *Coucal* (ASR 8) a Chanticleer-class submarine rescue ship. Ironically, Link perished five and a half years later in a submersible research vessel built by his father Edwin A. Link. The *Johnson Sea Link* became tangled in cable from USS *Fred T. Berry* (DD 858), a ship that had been sunk as an artificial reef. A decision to wait for Navy rescue vessels was disastrous when the low temperature levels reduced the efficacy of the CO2 scrubber that left Link and his colleague, Albert Dennison Stover, with dwindling oxygen levels leading to their death.

Now here you rest, you sorry old girl,
Bravo eighteen, Naval Shipyard Pearl.
Starboard you are to the pier alongside,
Four strong nylons keeping you tied.
All week long, they've carved out your vitals,
Surgeons called yardbirds, or some other title.
And so efficient have they been with the knife,
That only outside help will sustain you in life.
Fresh water, electric power, and steam,
Without them, you'd quickly die, it seems.
Around you lie other old gals in repair,
Paint stripped off and often left bare.
For they are ladies of the Pacific Fleet,
Temporarily down, but certainly not beat.
Many, like you, are well past their prime,
Having rendered services for a very long time,

........

And now you old B_______, I mean lovable dear,
Happy New Year to you in your twenty-fifth year.

# CHAPTER 4

## Late Night Laughter: Humor on the Midwatch

Thousands of sailors have honored the longstanding US Navy tradition of writing the midwatch log in verse. Some have made diligent efforts to compose serious poems, some have produced doggerel, and some have emulated the work of famous poets. Others, unsurprisingly, have incorporated humor in their efforts, gently poking fun at shipmates, superior officers, or members of the other services; complaining about drawing the midwatch duty; lamenting their absence from home; or even joking about required log information (e.g., location, course, ship status). Humor is not unique to the Navy, and it has a long history among military personnel[1] — a history that our midwatch poets continue to write. This chapter highlights a few of the efforts Officers of the Deck have made to wax humorous on their lonely midnight watch.

### Stuck with the Midwatch

The most frequently occurring theme for midwatch humor is the dismay associated with this double dose of bad luck:

standing the New Year watch and being expected to write a poem. Ensign J. B. Winningham summed it up well in this mournful entry in the midwatch log of the USS *Mullany* (DD 528):

> January first,
> of all the watches one could have
> this is by far the worst.
>
> Los Angeles is drunk tonight
> in liquor sweetly flowing.
> but on Mullany it's so quiet
> I can hear my fingernails growing.

Lt. j.g. B. C. Lorig seemed to border on despair as he did his midwatch duty aboard USS *McNair* (DD 679), writing

> Condition x-ray modified
> 'But I'm locked in', I softly cried.
> I wrote this verse before I died.

Occasionally, the OOD reports a sacrifice that goes beyond the typical boredom or lack of shore leave. Lt. j.g. E. P. de Merlier of USS *Forrest B. Royal* (DD 872), after questioning his own poetic prowess, reported what he sacrificed to stand the watch:

> But, Enough! I digress, this log may be an awful mess, unless my literary talents improve I'll be shining the X. O.'s shoes . . .
>
> Thus has Navy tradition been kept, tho' I broke a date to stand on this deck, but I hasten to add, err you think me a martyr, I've got the duty (She was the captain's daughter).

Other erstwhile poets avoid any flowery attempts at literary finesse and get straight to the point, as in this entry by Chief

Warrant Officer Donald L. Tupper on USS *Interpreter* (AGR 14):

> Present are many ships of the fleet
> And on many a quarterdeck stand frozen feet.
> Com Fair Alameda is SOP afloat
> And this is the damnedest log entry I ever wrote.

Lt. j.g. T. Umsted is, in his own way, equally clear as he closes his USS *Lester* (DE 1022) midwatch log with a simple plea:

> When Christmas comes December next
> To Santa I will write
> 'I'll take the watch most anytime,
> But please not New Years night.'

### Just the Facts

Sometimes the midwatch poet manages to strike a humorous note even while reporting the mundane facts of the ship's status. Such is the effort of Lt. j.g. H. H. Highstone as he recounts the situation aboard USS *Magoffin* (APA 199):

> Number 1 generator's putting out our power,
> But where do we get our bracing shower?
> From the pier! Fresh and salty waters flow,
> Telephone messages come and go;
> In fact we're getting everything but steam,
> But leisure, 'twould seem, is a steamy daydream.
> (This isn't quite true; we get no power either,
> But when the poet strikes, truth sometimes takes a breather).

Ensign J. D. Glenn manages to invoke not only a description of doubled lines, but also some double entendres as he logs the status of USS *Albany* (CG 10):

> Our port side is glamorous;
> The pier is feeling amorous.
> Her name is West Wall (with two L's);
> These Mayport girls are something else.
> We've used these same old lines before,
> But she keeps coming back for more.
> So now it's six lines doubled up—
> You'd think that that would be enough
> But no, she won't leave us alone, though
> We drink her water, use her phone.
> We give her all the trash we've got
> And use her for a parking lot.
> She takes it all without a fuss
> And uses words like 'we' and 'us.'

And aboard USS *Peregrine* (EMSP 373), Lt. j.g. A. J. Binsfeld combines an account of his ship's status with comments about his shipmates who are staggering back from celebrations ashore:

> The midnight moon gives a thirsty cheer
> The landlubber quaffs the new year's beer.
> Some old folks pout while youngsters kiss,
> And we on duty are missing this.
> Here we sit with our coffee grounds;
> Oh hell, I guess I'll make the rounds.
> The condition set is (modified) 'Yoke,'
> Machinery is working, nothin's broke.
> Moored at Delta-3, 'November' pier.
> Here comes the first man soaked in beer.
> Lines are secure, tide is slack;
> Ahoy! More drunks a'staggerin' back . . .

**The Ship**

Ensign J. D. Schneider writes from the viewpoint of his ship, USS *Hazelwood* (DD 531), to convey both basic log information and his sense of humor. He also manages to joke a bit about the Senior Officer Present Afloat.

> My skin is grey, my bottom's wet, I'm seventeen years old,
> And though I know my way around, my plant is very cold.
> Now here's a flash I just received; one I often get,
> A chicken laid an egg and said "Condition YOKE is set".
> Now a yoke's a yoke I always say, until it has been fried,
> Then I am sure Condition YOKE is slightly modified.
> SOPA here is COMDESLANT, king of many queens,
> [Commander Destroyer Force Atlantic]
>
> And I will utilize the fact when I learn what SOPA means.
> His home is the YOSEMITE, king of punts and prams,
> She's the ship that spies on me and belches nastygrams.
> Her hull is number D-19, I think that she's a ship,
> But that's an arbitrary guess—she's never left the slip . . .

Lt. j.g. D. C. Paulson also invokes the voice of his ship, USS *Fletcher* (DD 445), as he has some fun with his midwatch log:

> I'm close to her; she's really built, and also quite a talker
> I know her well, conceited as hell. Of course I mean the Walker

Six mooring lines around her, we can't move from this place
We're lying here at Baker two-five, Pearl Harbor, Naval Base
Receiving all services from the dock; for weeks we did not stir
T'was hoped her 'E's might rub on us, but we blew tubes on her
With all these 'she's' around us, you'd think it quite a treat
But how intimate can you get, with the U.S. Pacific Fleet

**Poking a Little Fun**

The OOD has, on occasion, used poetic license to direct (mostly) gentle jibes toward other officers and crew. In a series of limericks, Lt. j.g. C. E. Wright displays a sense of humor while observing his shipmates from USS *Lawrence County* (LST 887) as they make merry:

There was an old ship named Larry
Whose APE was frightfully hairy
The captain one night
Threw a party so bright
That the Snipe had more than he could carry . . .

There was a young rascal named Pel
In an Oriental bath he fell
But as ol' Pel leered
The girls giggled and disappeared
And the water turned cold as well.

Lt. H. R. Schmidt teases a litany of his mates on USS *Picto*r (AF 54), from the captain to the carpenter and machinist. For example:

Her Captain is a mustang [mustang – officer who began navy career as enlisted]
Who strums on his guitar.
He could never be a heel
Though he certainly is a tar.

The navigator is a portly guy
Who sees more stars than are in the sky.
He would much prefer to exchange this bore
For a lusty life with the pork chop corps.

Sometimes the jibes are a bit more pointed, as in this entry by Ensign C. L. Whitney of USS *Hanson* (DD 832):

I have the deck, Mr. Rowe's at the conn,
And the captain's on deck (but suppressing a yawn).
With the half-moon abeam at four minutes to three
I heard a short blast from the guide, Coral Sea.
'Watch out now, she's turning to starboard," I roared.
'No, No,' said the boatswain, 'The captain just snored.'

Lt. j.g. R. D. Fischl of USS *Dupont* (DD 941) is another OOD bold enough to not only note his skipper's night on the town, but also to detail the executive officer's New Year woes:

The XO so far is nowhere to be found
but the captain has returned after a night on the town.
At 0110 a rumor was rife
the XO had damn near lost his life.
It seems he had fallen under the attack
of a harbor water taxi, on the way back.
Having declined the Prairie's offer of treatment
he returned to the quarterdeck displaying deportment.
Accompanied by the faithful OPS boss
he claimed he felt like his 'fell off a hoss.'
We could tell by the events of the night

the new year was welcomed and brought in right!

At other times, the midwatch poet shows a semblance of empathy for his fellow crew members, as when Lt. j.g. D. S Kruger acknowledges the plight of his freezing USS *Cambria* (APA 36) shipmate:

From down on the pier come the phones and the water.
A pier sentry we must supply;
A cold job it's true;
Oh look! He's turned blue,
He's half frozen and ready to die.

Occasionally, shipmates require empathy for other reasons, as Ensign D. W. Buckingham notes in the log of USS *Frank Knox* (DDR 742) when a returning sailor interrupts his observance of the new year:

On the beach we can hear strains of 'Auld Lang Syne",
The toasts to the New Year with champagne and wine.
'Tis a time of rejoicing and one thing that's – Good Lord!
Messenger! Please, help that man crawl aboard.

Empathy is not always the order of the day when crew members stagger back to the ship, as Lt. R. K. Lehnus reported in the midwatch log of USS *Damato* (DDE 871):

Watch these fellows stumble toward the brow!
They make it aboard but I know not how.
What shall we do with our happy chum
Whose tummy is numb from so much rum?
'Put him in the scuppers, or maybe the hawsepipe!'
Said the J.O.O.D. to a bosun and snipe.

'Throw him in the whaleboat, or hang him by the yardarm!'
Yelled the C.D.O. in a tone of alarm.
And so amid this scene of good cheer,
Starts a happy new year at the 'D and S' piers.
At last that slow clock says quarter to four.
Our reliefs can shiver and we will snore.

Yet the midwatch OOD can at times be kind in warmly recognizing a fellow sailor, as Lt. j.g. P. B. Loomis does in this entry from the deck of USS *Fort Snelling* (LSD 30):

Just one more announcement
to make us all feel glad
at twenty thirty yesterday
Ensign THOMSON became a dad.

Her name is Farley Goodwin
weighed in at seven pounds eight
A brand new tax deduction
gives cause to celebrate.

Although it may not be widely known, US Marines have a long history of service aboard US Navy ships, even serving occasionally as OODs[2]. We might expect then that, like other shipmates, sea-going Marines would also find themselves the target of midwatch humor, as in this example by Ensign E. B. Bosler, OOD aboard USS *Boston* (CAG 1):

The Ninth Division's asleep in their racks
Dreaming of landings with M--"'s and packs.
They say they're Marines, but we know the truth
They tried to go Navy, can't read, so they goofed.

Most people probably do not associate humor with emergency situations. After all, what is funny about a crisis? However, scientists[3] have discussed the role of humor as a means of reducing stress, while enhancing the connections among people in an emergency. Unlikely as it may seem, Lt. J. O. Durand uses humor to describe a very real emergency on USS *Bluebird* (MSC 21). He tells a good story:

> Moored at Pier Uniform, starboard side to
> It's finally upon us, the night that we rue.
> In Charleston, all around this Naval Station
> People greet the New Year with capricious elation.
> While we receive from the pier services miscellaneous
> A crime is being plotted, both grotesque and quite heinous
> SOPA, COMINLANT, has devised us a lulu
> To save us from boredom—What else but Plan Zulu
> Plan Zulu—tonight! How can it be?
> As one the crew yells 'That old SOB!'
> What is the trouble? the RM is cursing
> Modulation is poor and the output is worsing
> Come on all you hearties—this is no joke
> 'But alas,' says the ET 'the transmitter is broke.'
> 'Smoke?' yells the gunner, 'In the radio shack?'
> 'Fire!' cries the Bosun, jumping out of the rack
> In a thrice a hose was there on the scene
> Quicker still on the valve the DC did lean
> 'No, No!' cried the OD 'This is radio gear.'
> But alack, through the din no one could hear
> When the water hit that SRT
> It lit up like a Christmas tree
> The ODs mind was quick to reason
> It was quite in keeping with the season
> The crew grew nervous, but this relax'em
> The OD repeated this old maxim
> 'Oh what a tangled web you weave
> When you practice Plan Zulu on New Year's Eve'

## Greeting the New Year

In one manner or another, midwatch poets celebrate, or at least acknowledge, the coming of the new year. Sometimes the new year is welcomed with a humorously begrudging tone, as in this entry by Lt. j.g. D. H. Swan of USS *Orion* (AS 18), whose watch is both dry and cold:

> The miscellaneous services we receive from the pier,
> Don't include what we'd want, like liquor or beer,
> Most of the crew are taking a snooze,
> While the Liberty party is out hitting the booze,
> So it's hail to the new year, farewell to the old,
> Now where's my relief, my damn feet are cold.

Ensign Robert Lotz, on the other hand, welcomes the new year aboard USS *Hugh Purvis* (DD 709) with a self-deprecating joke about his own skills as a poet:

> To end this verse as it grows
> we'd like to make this one thing clear
> to you and yours from us and ours
> have an ecstatically Happy New Year.
> Keats and Shakespeare would agree
> good poems should end in a couplet
> but though I try, for all my trouble it
> seems I muffle it and get a thruplet.

Meanwhile, Lt. j.g. R. L. Royalty imagines Lord Byron writing the log on a cold new year midwatch aboard USS *Pickaway* (APA 222):

> Various units of the Pacific Fleet,
> Yard and district craft too, noisily greet
> The New Year with blasts from their whistles and sirens
> And poetry better than George Gordon, Lord Byron's
> Could ever have been, had he composed his verse

On a cold quarterdeck at night; or worse,
If he had written the log of a ship: let's say
In the first four hours of New Year's Day . . .

Ensign C. E. Cole of USS *Alamo* (LSD 33) begins by welcoming the new year and finishes with one last poetic flourish:

Zero hundred to zero four
Usually our watchbill's biggest bore
But not tonight off Manila's shore
For the New Year's put his foot in the door . . .

On the quarterdeck I've stood my time
Composing this ungainly crime
Soon into my snug bed I'll climb
If I can find one last word to rhyme.

**Last Laugh**

These midwatch poems reflect the humor of sailors serving in a wide-ranging variety of circumstances in far-flung places around the world. Most are not raucous, rollicking jokes or word plays that cause us to laugh out loud. But they reflect the thoughts and sense of humor of junior officers honoring a naval tradition as they stand the watch in conditions that can be cold, lonely, and far from home. Humor may be best appreciated by those who understand the context from which it arises, and the New Year midwatch is surely a unique context. Besides missing the New Year frivolity enjoyed by their shipmates, those who stand this watch are also missing holidays at home, significant others, and even the birth of their own children.

Good humor includes the ability to laugh at oneself[4], and a number of our midwatch poets show that ability, as they make light of their poetic talent or their misfortune in drawing the midnight duty. Many also demonstrate the ability to laugh with their fellow sailors, with jokes and jibes that are light-hearted and good-natured. They manage this against a backdrop that includes the Cold War and the conflict in Vietnam, and in ports as distant and diverse as Europe, the Philippines, and Japan, as well as numerous U.S. stations. And now, a half century or more later, they can still make us smile.

# SIDELIGHT 5

## Lost, but not Forgotten

While many deck log poems share similar content and meter, occasionally one emerges with content like no other. Such is the case of the 1968 New Year's Day midwatch aboard USS *Robert A. Owens* (DD 827) written by Lt. R. K. Bertolotti. Bertolotti presents the reader with a mystery about a missing ship.

> I am here to investigate the improbable loss
> Of a ship of the line without good cause.
> She was safe on the blocks of Drydock four
> On New Year's Eve and then - - no more.

The poet continues with a litany of statements regarding the ship's condition after which he claims to have figured out the mystery of the missing ship. Unfortunately, for the reader, the answer is in code. The reader must decipher the message using the poet's "key" that relies upon specified letters scattered throughout the poem. The first part of his "key" begins with:

> For assistance in tracing the impossible loss
> Of a ship of the line, without good cause

A flash of brilliance my sight impaired
Revealed the answer, the case was cleared.
Letter one, line two, word seven for a start.
Last letter, line three, well deduced, quite a start.
Line five, fourth word, letters two and three
Then on to the last word, letter before e.

The final answer is "Wrong Dock."

Tack on a K and start out without fear
For certainly Owens always was here.
Baffled by docks, an identical pair
Is a mysterious way to break in the New Year.

# CHAPTER 5

## Duck and Cover: The Threat of Annihilation in Cold War Deck Log Poems

> Last year saw aggression and war and strife
> With "peaceful" nations struggling for life.
> Spies in the sky, atom bombs all around,
> The populace taught how to hide underground.

Americans who grew up in the Cold War years remember all too well the nuclear drills practiced regularly each school year in response to the threat of an atomic attack. After the Soviets successfully tested a bomb in 1949, President Truman sponsored the production of the film *Duck and Cover* in 1951 to be shown in schools across the country. In the film, Bert the Turtle demonstrated the safety technique of dropping to the ground and covering your head and neck (Bert retreated into his shell.) A whole generation of young students, from elementary grades through high school, were shaped by regularly acting out a doomsday scenario, diving under their desks and covering. In the poem excerpt above from the log of *Sea Devil* (AGSS 400) in 1961, Lt. j.g. P. A. Peterson seemed to lament what the Cold War world had become. Twenty-two months after Peterson wrote his lament, the doomsday threat would knock on America's door in the Cuban Missile Crisis, when President John F. Kennedy drew a line in the sands of Cuba and the world held its breath as the Soviet Union's Nikita Khruschev decided the next move.

The jaunty music and light-hearted approach of *Duck and Cover* did little to assuage the fears of school children or of an American public who had come to realize that the weapons used by the U.S. at Hiroshima and Nagasaki in 1945 could now be used against their own country. As Sarah Pruitt said in 2019, "The now-infamous duck-and-cover drills simulated what should be done in case of an atomic attack—and channeled a growing panic over an escalating arms race."[1] Not unexpectedly, that angst was felt by young officers on midwatch duty, and their poems evoked the growing struggle against Communism.

The images of cities decimated at Hiroshima and Nagasaki as a consequence of the dropping of a single bomb stirred the most fundamental fear in the American psyche, the fear of annihilation at the hands of Russia. Aboard USS *Halibut* (SSG (N) 587) in 1962, Lt. P. V. H. L. Duckett powerfully expresses the dread felt by so many.

> Each man hopes as the days transpire, it won't be our job to unleash Hell's fire,
>
> Hoping that no one sounds the knell that will turn God's Green Earth into a flaming Hell.

That same year a young Lt. j.g., C. M. Kennedy, the OOD aboard *Wahkiakum County* (LST 1162), ends his poem by expressing the stark alternatives that lay before the world.

> New Year, New Year, what do you hold,
> A future bright, or more of the old?
> Will man find peace, or peace find man,
> Will the hydrogen bomb destroy our land?

The fear of Communism had been swelling in the Western mind since early in the twentieth century. Following the 1917 Bolshevik Revolution in Russia, the common descriptor "Reds" as applied to Communism became popular, reflecting the red flag carried by the revolutionaries. One of the sources of the term's popularity in American speech was an article written by Attorney General A. Mitchell Palmer in 1920 titled "The Case Against the 'Reds.'"[2] And the American response to early Communist activity in the country has come to be known as "The First Red Scare" (1919-1920). Russia's rapid rise to power following World War II amplified the fear. The term "Reds" has had a long half-life, still used a hundred years later, and it appeared often in deck log poems written during the Cold War.

Lt. H. W. Walters, in his 1956 deck log aboard USS *Salisbury Sound* (AV 13), uses it when referring to the ship's role in the struggle as part of West Pac,

> And by our presence keep the Reds in check
> (Assisted by the carriers and maybe Chiang Kai Shek).

Lt. j.g. P. H. Herzing, writing the first log of the year 1957 aboard USS *John W. Weeks* (DD 701), describes Navy ships far from home protecting the country "From perils and threats of Red tyranny." On New Year's Eve of 1962, the duty officer aboard USS *Spikefish* (SS 404), Lt. j.g. James R. Burnett, refers repeatedly to the Communist threat as the "menace red." The following verse of his poem is representative of the whole. To read the entire poem, see Chapter 10.

> So set the watch my shipmates bold
> And keep it taunt and well
> And squarely face the menace red
> This excrement of hell.

Burnett then closes his poem with a stanza encouraging resolute duty to keep the menace from our land. The New Year's log on USS *Caddo Parish* (LST 515) in 1964 also offered an admonition for men of the Navy to stand strong in their duty to maintain freedom. Signalman 3rd Class D. J. Sanders and Lt. j.g. F. B. Wheeler, Jr., expand the focus of their warning, asserting America's responsibility for stopping the Communist threat throughout the world.

> The amphibious base remains alert tonight.
> To defend our freedom from the Communist might.
>
> . . . .
>
> We have seen what has happened in Havana and Berlin
> And our job is to keep it from happening again.
>
> . . . .
>
> Throughout the world this very night
> The Reds continue their endless fight.
> And they will not stop until the entire world
> Has seen their communist flag unfurled.

The spread of Russian influence around the world was a concern in many deck log poems during the Cold War years. Lt. P. W. Steinhagen addressed an opinion shared by some that the atomic age made the Navy obsolete. His New Year's deck log in 1962 aboard USS *Boxer* (LPH 4) refers to the hammer and sickle on the flag of the Great Bear of Russia, claiming that the enemy understands the value of Naval strength.

> For those whom the Fleet is obsolete,
> As shield for hammer-blows:
> See the great bear rear across the seas,
> He sure as Hades knows.

The Great Bear's appetite seemed unquenchable as it devoured country after country in the East, and then, with Fidel Castro's rise to power in Cuba after the fall of the Batista government in December 1958, it threatened America's shores. In April 1961, the U.S. backed an invasion of the island at the Bay of Pigs, attempting to overthrow the communist government, but the effort failed miserably. Lt. George B. Rufho imitating the structure of the popular poem "The Night Before Christmas" in his 1961 deck log from USS *Greenfish* (SS 351), excoriated Russia's attack on Democracy and the widespread creation of Russian satellite countries. After claiming that "Russia in the East and the US right here / Cannot come to terms even over a beer," he addresses the real threat.

> And what to our wondering eyes should appear
> But a sky full of missiles and atomic war gear.
>
> With a little old driver so bitter and quick
> That I knew in a moment it must be Khrush-nick.
>
> More rapid than eagles his satellites came,
> While he whistled and shouted and called them by name.
>
> Now Cuba – now China Poland and Rumania
> On Hungary – Germany – Bulgaria and Lithuania

Ensign L. G. McIntyre, OOD aboard USS *Warbler* (MSC 206) in 1962, a mere eight months after the Bay of Pigs calamity, lamented the passing of the old days of sailing with the sextant and the stars and its being replaced by the atomic age.

> The minute atom drives monsters of steel,
> With scram rods and meters and buttons to feel,
> Replacing mad fires and strong muscled men,
> With men high in knowledge of roentgens, I ken.

McIntire warned of Russia's determination to replace freedom and democracy with fear and communism, and he decried the many current conflicts created by the global struggle, including "communist Deutschland tearing Berlin" and the looming war in Vietnam. The verse below addresses the growing struggle in Cuba, using the sickle from Russia's flag as a metaphor for Russian itself.

> The sickle has Cuba at its foremost part,
> Driving toward our healthy heart,
> But we are strong, and she we will rout,
> Like a cub in a camp, we will throw her out.

With this optimistic prediction coming just eight months after the Bay of Pigs, a reader might understandably question McIntire's optimism, though President Kennedy's success in the Cuban Missile Crisis nine months later, in October, would perhaps justify it. In January of 1963, just two-and-a-half months *after* the Cuban Missile Crisis, Ensign Robert W, Beerkircher, writing his log on the oiler USS *Elokomen* (AO 55), referred to his ship's participation in the Cuban blockade as the "strangling" of Fidel Castro.

> Into port and out again, prepared to stand against the ominous foe
>
> The Elo south straightway her course did make to strangle Fidel Castro

He then expressed pride in the ship's work and tells of having visual contact with a Russian submarine.

> Hours on end she supplied the Fleet with fuels, cargo and mail
>
> Night-fueled a ship in visual range of a Russian sub's phosphorescent trail.

The opening lines of the log for USS *Lester* (DE 1022) in 1962, written by Lt. j.g. H. Bolton Finn III, seemed to reflect a growing pessimism about the future.

> At this coming of the "bright" new year,
> On this earth there is not much true cheer.
> In young Castro's land, Berlin and Viet Nam,
> Violent men speak of the world to come anon.
> While throughout the US all try to forget,
> The fact remains we are facing a threat.

That threat had been expanding. Development and possession of nuclear weapons had been proliferating. Great Britain was closely involved in the US development of nuclear weapons in the Manhattan Project, notably after the Quebec Agreement in August 1943 between Franklin D. Roosevelt and Winston Churchill, and succeeded in their first independent testing in 1952, the same year as Russia. In 1960, France became a nuclear power, and four years later, on October 16, 1964, China successfully tested the bomb. Two-and-a-half months after China's entry into the nuclear club, Lt. j.g. Wiley G. Grantham, serving aboard USS *Ault* (DD 698) 1965, noted the expanding threat.

> The whole world is torn with strife and dissension.
> The Congo is boiling with internal tension.
> Soldiers are dying in South Viet Nam,
> And China exploded a nuclear bomb.

Throughout the Cold War, despite a clear recognition of the imminent threat of annihilation in nuclear war, young American officers in New Year's deck logs never lost their enduring sense of humor. In addition to courage and hope, humor may have represented the perseverance of the American spirit better than any other characteristic. As they had always done in war and in peace, the OOD's on US Navy ships brought to the grim reality of the global situation an uncompromising American humor. MCCS W. C. Bluemling, writing on New Years in 1965 aboard USS *Fulton* (AS 11), mocks Fidel Castro with a pun on an old home remedy that was so bad it was sometimes used as punishment for children.

> Now Cuba I'm sure cannot long endure
> Being used as a communist foil,
> It will be no surprise when Cubans realize
> The results of too much Castro oil.

Unfortunately, Bluemling's prophecy of the fall of communism in Cuba, like McIntire's prediction in 1962, was not as successful as his pun. The Castros held power in Cuba for fifty-four years, and the island is still ruled by a communist regime in 2026.

Not all the humor in the Cold War deck log poems was aimed directly at the communists. Lt. j.g. W. E. Mosier mentions one of the great accomplishments of the age in his 1956 poem aboard USS *Nantahala* (AO 60).

> The last year saw some ups and downs
> With times both fat and lean,
> The market went up, the Summit came down
> And the polio vaccine.

Though Mosier mentions the Kremlin in the next stanza of his poem, his humor is aimed rather at the great American pastime

> In addition to this was a Middle East threat
> And the usual Kremlin queries,
> Memorable too is this was the year
> That the Yankees lost the series.

**Conclusion**

Though Russia's sinister aggression and the threat of atomic annihilation were never far from the minds of young American Naval officers, their optimism and hope were also there. Lt. K. A. Porter, aboard USS *Medregal* (SS 480) in 1959, expresses it clearly. After saying the end of the world did not come as forecast, he offers advances in science, space, and American culture, assuring the reader that "the Russians won't swarm us."

> Yet history may find that the largest effect
> Judged many years hence when the view is select
> Is not that of merchandise, glittering and neat,
> Nor even our missiles, nor nuclear fleet,
> But rather the growth in the minds of us all
> Of the seeds of a movement of stature so tall.

He then couches America's hopes for greatness in terms of the highest moral aspirations of the country.

> Morality, spirit, intensity, honor,
> Though technically strong, without we're a goner.
> Though it seems like a stalemate in arms is before us,
> This kindling of chances for greatness can soar us.

# SIDELIGHT 6

## Give Me Liberty………

As always we must, on the hardships to come,
With a spirit of Trust and "Get the Job Done."
We each did tremble, as her colors were raised,
With pride for the symbol we all have praised –
And vow to add fame, through duties done well,
To its historic name – the *Liberty* Bell.

So ends the first deck log entry for USS *Liberty* (AGTR-5) as she was commissioned on December 30, 1964. It espoused the "can do" attitude of a newly assembled crew to handle tough times that may lie ahead. Lt. Lester A. Morcerf, Jr., penned the rhyme for this old "Victory" ship that began service as SS *Simmons Victory* at the end of WWII. She joined the Navy as an AGTR – Auxiliary, General, Technical Research ship. It was no secret, however, that this type of ship listened to electronic communications of other countries and communicated them back to the US. She was a "spy ship" and fully supported the activities of the NSA (National Security Agency).[1]

The Cold War was in full swing. With USSR and China vying for influence around the world, the US used electronic signals intelligence to keep track of their machinations. Proxy wars broke out periodically, and ships such as *Liberty* often deployed nearby. In her early deployments, she spent time off the west coast of Africa. She went back to the US for refitting in late 1965. While in Norfolk on the first of January 1966, Ensign John McCorkle wrote the midwatch poem for the log. He finished his entry with an excerpt from the Navy Hymn:

> Eternal Father, Strong to Save,
> Whose arms hath bound the restless wave
> who bids the mighty ocean deep
> its own eternal limits keep
> Oh hear us when we cry to thee
> For those in peril on the sea.

Ensign David G. Lucas' 1967 New Years midwatch poem expressed hope for a happy new year, and the lyrics of the Navy Hymn became all too prophetic for *Liberty*. She and her crew would know "peril on the sea" first-hand.

On June 2, 1967, *Liberty* stood out from Rota, Spain, and sailed into the Mediterranean Sea to collect intelligence. Tensions in the Middle East involving Israel, Syria, Jordan, United Arab Republic (Egypt), and others reached a boiling point. The US, USSR, and China all had stakes in the outcome of any hostilities. On June 5 fighting broke out between Israel and Syria, Jordan, and Egypt. The Six Day War began while *Liberty* sailed in international waters north of the Sinai Peninsula. The Happy New Year that Ensign Lucas wished for *Liberty* abruptly came to an end on June 8 when Israeli aircraft and motor torpedo boats attacked her. She and her crew were in a fight for their lives, and Ensign David G. Lucas was caught up in the middle of the fray.

Just five days after the attack Lucas was called as a witness in an inquiry into the incident. In fact, the official record stated, “Let the record show that this witness is being called out of order that he may be medically evacuated to a hospital ashore for [deleted description of his wounds] and that he is experiencing considerable physical discomfort at this time.”[2]

Lucas described being in the wardroom and hearing the first attack and believing that the motor whale boat had fallen from its davits. As he headed topside, GQ (general quarters) were sounded as a second wave of planes hit *Liberty* requiring him to “hit the deck” while shrapnel splattered around him. Lucas went back to the wardroom to check on casualties and found that the ship’s executive officer, Lt. Cmdr. Philip Armstrong, was in very bad shape. Making his way up to the pilothouse, Lucas found the captain, Cmdr. William McGonagle, Seaman John La Mar, who was lying on the deck, and Quartermaster 3rd Class Francis Brown who was manning the helm.

It was at this point that Liberty took a torpedo hit and began a list to about ten degrees. Motor torpedo boats got close enough to see identifying details and the captain appointed Lucas an impromptu quartermaster with the command, “Log it!” There being no quartermaster available at the time, Lucas continued to make entries into the log for the next 45 minutes.

During another attack the captain warned the three others in the pilothouse to take cover. Quoting from Lucas’ testimony,

> Everyone took cover as best we could. There were shells, or fragments of shells, flying in through the starboard hatch, which no one had been able to secure. A quartermaster [Brown], when these shells started flying, stepped back from the helm to get as much protection from the corner of the chart room as possible. He still had his hand, one hand on the wheel. I was two paces to his left, and two paces

behind him. A fragment hit him, I think from behind. It must have come through the bulkhead in the chart room. He let out a gasp, fell backwards into the chartroom, and within say a minute, was dead.

The remainder of Lucas' testimony described how he took the helm and began steering the ship. He also told of looking out at gun mounts only to see the torn bodies of shipmates. Lucus and Captain McGonagle were both wounded in the attacks. McGonagle continued to conn the ship even though at least one of his wounds required a tourniquet due to "profuse bleeding." According to Lucas, the captain "insisted on being everywhere that he could. He was on both wings, in and out of the pilothouse, taking pictures." At one point, Lucas was asked if the captain was requested by the doctor to leave the bridge. He responded, "…it would have taken ten people the doctor's size to even begin to get him [McGonagle] off the bridge." In fact, the captain stayed in direct command on the bridge for 17 hours before allowing himself to be relieved.

When the attack was over, *Liberty* had paid a huge price.[3]
Officers: Wounded in action 11; killed in action 3
Chief Petty Officers: Wounded in action 4; killed in action 2
White hats: Wounded in action 156; killed in action 26
Marines: Wounded in action 1; killed in action 2
Civilians Wounded in action 1; killed in action 1
Total Wounded in action 173; killed in action 34

Many decorations were awarded to the crew, including a Medal of Honor, two Navy Crosses and 12 Silver Stars (the three top Navy awards for valor under fire). Two hundred and eight received Purple Heart Medals for suffering wounds. Lucas earned the Silver Star. McGonagle received the Medal of Honor. It was presented to him in secret by the Secretary of the Navy at the Washington Navy Yard rather than at the Whitehouse by the President.[4]

The Israelis claimed that they attacked *Liberty* by mistake believing it to be an Egyptian ship. They apologized and eventually paid over 90 million dollars [in 2025 dollars] in compensation. Many analysts still believe that the attack was not an accident but rather an attempt by Israel to stop the US from eavesdropping on their activities.

No matter the ultimate cause(s), *Liberty* was considered too damaged to repair, and she was sold for scrap in 1973.

There would be no more midwatch poetry entered into *Liberty's* log.

# CHAPTER 6

## To Sea or Not to Sea? That's a Good Question

The longstanding U.S. Navy tradition of writing the New Year midwatch log in verse has produced thousands of poets. Some of these budding bards have struggled mightily, producing tortured rhymes, humorous complaints, and laments about their ill-fated lot as unwilling writers. Some, however, have proven able and creative, honoring the tradition with good will and enthusiasm, often producing credible original poems that bring a smile to the reader's face. Still others have undertaken the task with an eye to the precedents set by well-known poets of the past, or forms of verse that are widely familiar, as they mimic writers both classical and popular.

Review of ten thousand or more midwatch poems found a wide variety of attempts to follow the work of famous writers while also including such information as location, course, condition of readiness, and sources of power. Thus, young officers of the deck have written in the vein of such luminaries as William Shakespeare, Edgar Allan Poe, Lewis Carroll, T. S. Eliot, Robert Frost, Rudyard Kipling, and more.

Their poems are sometimes humorous, sometimes solemn, and always interesting. And they achieved this against the backdrop of the Cold War and the conflict in Vietnam. Whether the poet apologizes for his literary ineptitude, or waxes eloquent in flowery language, each poem brings a unique perspective as the writer works to incorporate the requisite log content while maintaining the rhythm and structure of a well-known model.

**Clement Clarke Moore (1779-1863)**

The most frequently copied poem in midwatch logs is Clement Moore's *A Visit from St. Nicholas*, better known as *The Night Before Christmas*. This should not be surprising; the New Year's midwatch follows on the heels of Christmas, with many of the logs referring to Christmas lights, on ships and ashore. Nearly all the midwatch variants of this poem begin very much as Moore did, and as Ensign J. Lynch wrote while serving on USS *Georgetown* (AGTR 2) on January 1, 1966:

> T'was the night before New Year's and all through the ship
> Not a sailor was stirring on the pier or the slip

And, like the children in Moore's poem, sailors too have their dreams, as Lt. j.g. F. B. Schoenberger observed in 1957, aboard USS *Oglethorpe* (AKA 100):

> The salts were all snuggled tight in their beds
> While visions of sweethearts danced through their heads.

Although many ships were in port as the new year arrived, USS *Newell* (DER 322) was at sea on New Year's Day, 1957, when Lt. j.g. W. H. Fleak described his ship's progress:

One-eight-zero's our course, we're Pearl Harbor bound.
She leaps between waves like an old greyhound.

Of course, he who stands the New Year watch observes many things, including the return of shipmates who have stayed a little too long in the pub. Ensign H. A. Johnson described one such sailor as he found his way back to USS *Gunston Hall* (LSD 5) after ringing in1964:

I gazed down pier eight
to see what was there,
'twas a small fat sailor
shaped like a pear . . .

He staggered, he wobbled,
he stumbled, he fell,
but by the tone of his laughter
we knew all was well

Many poets brought their version of this poem to a close with a traditional wish for a happy new year, while some gave it a humorous flourish. Writing from USS *Pickaway* (APA 222) on January 1, 1962, Ensign H. B. Parkhurst was one of those finishing his log in the traditional way:

No more is required; no more shall I write,
So Happy New Year to all, and to all a good night.

**Edgar Allan Poe (1809-1849)**

Edgar Allan Poe is another author frequently emulated by midwatch poets, virtually all of them producing their version of *The Raven*. Perhaps it is fitting that those standing the midnight watch would be intrigued by the eerie, haunting feel of Poe's poem. Poe's opening was mirrored by Lt. j.g. T.

J. Ready III, OOD on USS *Halsey Powell* (DD 686), when he wrote the midwatch log in 1962:

> Once upon a midwatch dreary, as I checked, all weak and weary
> Over many a quaint and curious sailor from ashore
> . . .

Lt. R. E. Burton opened the 1969 midwatch log aboard USS *Sargo* (SSN 583) with a lament about drawing the midnight duty, and finished his first stanza with Poe's trademark "Nevermore:"

> Upon a New Years evening dreary
> While I ponder weak and weary
> With wife and friends at parties drinking
> Here I sit with spirit sinking
> New Years Duty Nevermore

The atmosphere engendered by *The Raven* was captured by Ensign W. T. Dannheim, standing the new year's midwatch on USS *Mitscher* (DL 2) in 1964, while also skillfully incorporating essential log information:

> YOKE is set in all the spaces, checks are made in many places,
> All is well throughout the ship and readiness is one plus four.
> Our Christmas lights are burning clear from power generated here,
> We use our own though help is near and could be taken from the shore.
> We generate our own and do not take our power from the shore.
> This and Steam and nothing more.

In 1970, Ensign R. T. Parette proved similarly capable as he managed to include a wealth of ship's data in this stanza from the log of USS *John Paul Jones* (DDG 32):

> Ah, distinctly I remember, on the last day of December,
> and each separate bored-stiff member of the crew had hit the rack
> Long Beach shipyard was our berth place, doubled mooring lines our girth braced
> through our shipyard work we now raced
> (our boiler casings had been cracked)
> Sundry yardcraft, PACFLT vessels wait with readiness we lack

As he came to the end of his 1962 rendition of *The Raven*, Lt. j.g. J. D. Coolbaugh wrapped up the midwatch log on USS *Barry* (DD 933) with a bit of self-deprecating humor and a bow in the direction of Mr. Poe:

> While my ode is now quite nigh completed,
> My hoard of homonyms depleted,
> There is one thought that, in all fairness, now I must surely show.
> I should like to take this chance to send
> My apologies and make amends
> To a bard with far more talent than I, Mr. Echo Alfa Poe.

## Henry Wadsworth Longfellow (1807-1882)

Like Moore and Poe, Henry Wadsworth Longfellow was a popular nineteenth century poet, and numerous midwatch logs reflect the influence of his well-known *Paul Revere's Ride*. Ensign L. W. Allan, for example, began his 1964 midwatch entry aboard USS *Cavalier* (APA 37) in a way familiar to anyone who has read Longfellow:

Listen my shipmates and you shall hear
The midnight log of the CAVALIER.

Ensign C. Kollmeyer reported on the condition of his USS *Hopewell* (DD 681) shipmates returning from New Year's festivities:

T'was New Year's Day in '63
The men were staggering back from a spree,
and most were looking not quite well.

And Ensign E. C. Coffey of USS *Edson* (DD 946) could not have imagined that we would be reading his midwatch entry sixty years later, as he acknowledged both the arrival of the new year and belief in the obscurity of his log:

It is the first of January in '65
and perhaps but one is now alive
who will remember this famous day and year,
and the midwatch entry at the sixteenth pier.

**Walt Whitman (1819-1892)**

Although Walt Whitman's best-known poem may be *Leaves of Grass*, the choice of several midwatch poets was his *O Captain! My Captain!* Writing in 1865, Whitman used the metaphor of the death of a ship's captain as he lamented the assassination of President Abraham Lincoln. This metaphor clearly struck a chord with several midwatch poets, including Ensign W. R. Illing of USS *Impervious* (MSO 449), in 1970:

O Captain, My Captain
Our trip has not begun . . .

His ship was in port, but Ensign Illing clearly longed to be at sea:

> O Captain, My Captain
> Please hurry if you can
> O bring me back to the ocean deep
> I'll see no more of land.

And as 1961 dawned, Ensign Alan S. Krech gave Whitman's poem a humorous twist that bordered on the sardonic as he described his shipmates on USS *Sagacity* (MSO 469):

> But wait! We have our new exec, ex boatswain mate is he.
> I'm sure that with his knowledge we'll not sink into the sea.
> It's just that when we need him most, he can not leave the rail
> For he's been fishing all year long and finally caught a whale.
> It's J.O.s to the rescue. Let's see what they have mastered.
> The 1st Lieutenant is home in bed. The Engineer is plastered.
> The Ops Officer can not help, he's almost bound to swamp us.
> For 'though he made it to the bridge, he can not reach the compass.

We can only wonder what Whitman would have made of these seafaring efforts to follow his poetic lead!

## Lewis Carroll (1832-1898)

Lewis Carroll, whose fame is tied to *Alice's Adventures in Wonderland*, was also a poet, and several sailors chose to write a play on his *The Walrus and the Carpenter*. Lt. j.g. L. S. Bushnell, OOD on USS *Macon* (CA 132), toyed with Carroll's opening line in this 1957 midwatch log entry:

> The sun wasn't shining on the sea, shining with all its might;
> as we were anchored in berth Foxtrot (Cannes, France) to everyones delight.

Lt. R. O. Darby III opened his 1969 log on USS *Robert H. McCard* (DD 822) with a parody of the most famous line from Carroll's poem:

> The time has come this Ensign says
> to talk of many things;
> of ships and seas and sealing glands
> and after bearing rings.

Carroll's poem *You Are Old, Father William* also makes an appearance, in a clever log penned by Ensign P. N. French of USS *Floyd B. Parks* (DD 884). On January 1, 1968, after beginning with apologies to Carroll, he wrote:

> 'You are old, Father Time,' the O.O.D. said,
> 'and your hair has become very white;
> Although you would rather be tucked in your bed,
> Please help write the mid-log tonight.'
>
> 'Oh come,' said the spirit, 'that's not such a chore;
> the writing of logs takes no brain.
> To set the ship's status to lyrical verse
> is more of a joy than a pain.'

**John Masefield (1878-1967)**

In 1902, at the dawn of the twentieth century, John Masefield wrote *Sea-Fever*, with its famous opening line "I must go

down to the seas again . . ." Masefield expressed wanderlust and a love of the sea. Sometimes, with their ships in drydock or moored far from home, Navy poets have also dreamed of once more going asea. Writing in Masefield's meter, and welcoming 1965, Lt. B. S. Weisman said of USS *Chicago* (CG 11):

> A challenging year for us starts today
> fulfilling our vital mission,
> for cruisermen always lead the way,
> according to naval tradition.

Then, just as Masefield asked only for a merry yarn from a fellow rover, Lt. j.g. David A. Wagner, of USS P*onchatoula* (AO 148), writing sixty-four years later, made a similar request for his ship:

> And all we ask is the comradeship of our daring fellow-rovers,
> And a sweet sleep and a tall beer, when the long trip's over.

**Robert Service (1874-1958)**

Soon after the appearance of Masefield's popular poem, two of Robert Service's works became big hits. In 1907, he published both *The Cremation of Sam McGee* and *The Shooting of Dan McGrew* and went on to become known as the Bard of the Yukon. The lyrical rhyming pattern of his poems made Service both popular and memorable—facts not lost on some of our midwatch poets. For example, Ensign J. M. McFadden launched his 1962 USS *Robinson* (DD 562) midwatch log in typical bold Service-like style:

> There are strange things done in the midnight sun by they that

steam the Atlantic.
Daring and bold, & perhaps a bit cold, but albeit not the
least frantic,
Calm & collected, & never dejected, the ones that put out to sea
May tend to grieve, for it's New Year's Eve, & what a hell of
a place to be.

Two years later, aboard USS *Durant* (DER 389), Ensign T. Fisher III carried on with Service's swashbuckling style:

There are strange things done in the midnight sun
by sailors both brave and bold;
The Aleutian swells have their secret Hells
That would make your blood run cold;
The Northern lights have seen queer sights,
But the queerest they do ever see
Are the ship's that roll on Barrier Patrol
Far out in the North Pac Sea.

When USS *Porterfield* (DD 682) was deployed from the Philippines to the South China sea at the beginning of 1961, Lt. j.g. S. T. Webster wrote this in the midwatch log (The terms "whooshers" and "snorters" probably refer, respectively, to jet- and propellor-powered aircraft):

With Task Group Seventy-Seven point Seven,
We're scheduled to watch and assist,
Whooshers and Snorters,
With engine disorders,
that try for the carrier but miss.

And Ensign J. R. McAlpin closed out his log on USS *Pinnacle* (MSO 462) with a brief account of the crew's New Year merry-making as they celebrated the arrival of 1966:

This was the scene on New Years' Eve,
As gathered one and all,
To drink and yell with merriment,
To stagger but not fall.

### A Medley

Although most Navy poets, as they attempted to mimic the work of famous writers, picked a single model to emulate, Ensign R. J. Graves took a different tack, making his midwatch log aboard USS *Forrest Sherman* (DD 931) a medley of easily recognizable styles. Some of his lines from New Year's Day, 1962:

Once upon a midnight dreary,
As the FORREST SHERMAN sits so weary . . . [Poe]

We think that we shall never see
The SHERMAN carried off to sea . . . [Joyce Kilmer]

Africa was lovely, dark and deep
But we had promises at home to keep . . . [Robert Frost]

Enemies, allies, and countrymen:
Lend me your ears . . . [Wm. Shakespeare]

Out out brief SOLANT Amity Three
Back to twist around the Christmas tree . . . [Wm. Shakespeare]

Merry merry white and airy . . . [Mother Goose]

Other poets received attention in the midwatch logs of the 1950s, 60s, and 70s, sometimes in apologies, and sometimes in self-deprecating comparisons. Among them were William Blake, Robert Louis Stevenson, Lord Byron, Alexander Pope, John Keats, and Carl Sandburg. Still others were mentioned, though not emulated by the midwatch poets. These included Ralph Waldo Emerson, Henry David Thoreau, Percy Shelley, John Milton, and Omar Khayam. Interestingly, references to female poets are notably absent in the logs of this era.

**Verse Forms**

A few of our midwatch poets emulated not a famous poet, but a particular form of verse or song. Thus, we have songs, limericks, poems modeled after those of the beat generation, and even a take-off on an old bit of bathroom graffiti doggerel.

*Songs.* In addition to the Viet Cong, *Puff, the Magic Dragon*, composed in 1962 by Peter Yarrow and based on a poem by Leonard Lipton, was on the mind of Ensign J. M. Fuchs when he wrote this 1966 entry aboard USS *Paricutin* (AE 18):

> Puff the magic AE
> lives down on the sea
> and frolics in the morning mist
> with the ships of the seventh fleet
>
> Little Victor Charlie
> was hiding in the grass
> when a Navy jet dropped one of Puff's bombs
> and knocked him on his . . .

Although the ensign did not complete the sentence, we can infer the word he omitted!

Lt. E. L. Therrien, an English major from College of the Holy Cross, seemed to be in a patriotic mood when, on January 1, 1969, he extolled the virtues of USS *Thomas J. Gary* (DER 326) to the tune of the *Battle Hymn of the Republic*:

> Mine eyes have seen the Gary at the pier in old Key West.
> She is port side to Pier Delta Two, but not within a nest.
> She's all alone with doubled lines, her diesel plant at rest.
> 'T. J.' is steaming on!
> Gary, Gary Hallelujah,
> Gary, Gary Hallelujah,
> Gary, Gary Hallelujah,
> T. J. is steaming on!

And we can only assume that Lt. j.g. E. L. Brasseur was a fan of Gilbert and Sullivan, or at least of the Broadway musical genre, when he welcomed the year 1958 from the East African port of Djibouti while aboard USS *Hyman* (DD 732). This is his rendition of the classic *Modern Major General* from *The Pirates of Penzance*:

> This is the very model of a modern naval man-o'-war;
> We're sitting in Djibouti, where the SOPA is our Commodore.
>
> Our name's the Happy HYMAN, and our number's seven-thirty-two;
> The Commodore's aboard now, though he sometimes rides the Purdy, too.

Lt. Brasseur signed off his new year midwatch with a bang:

> This is the very model of a modern naval man-o'-war.

*Limericks.* Limericks have long been a favorite form of verse with many, including some of our young Navy poets. Ensign R. P. Onorati of USS *Charles S. Sperry* (DD 697), for example, was sufficiently enamored with the limerick to make his midwatch log a string of ten of them on the first of January 1962. He began in this way:

My poetry is really a crime,
but Navy custom requires a rhyme,
so with vigor and zest
I embark on this test,
for this midwatch I've nothing but time.

Lt. S. T. Webster seemed to be referring to a joke or incident known only to crew of USS *Bang* (SS 385) when he wrote this rather foreboding introduction to 1964:

If your chief of the Boat is 'The Greaser'
And the crew members mangle his sneezer
Get him off of the ship
If you go on a trip
Or you'll bring him back home in the freezer.

Lt. j.g. Glenn James Secrest showed pride in his submarine, USS *Jallao* (SS 368), as he closed out his 1966 midwatch log (The term "VAN" here refers to "advanced guard"—vanguard):

Some say there's no vision for man,
Like destroyers awash in the VAN;
But our sub's a front-runner,
And we're hard to uncover,
So we'll challenge any tin-can.

*Beat Poems.* The late 1950s and early 1960s were the era of the Beat Generation, including such writers as Allen Ginsberg and Jack Kerouac. Writers of this period often rejected traditional narratives and questioned cultural traditions. As we might expect, the influence of the Beat Generation can be seen in a few midwatch poems. We include two examples here.

Lt. j.g. W. E. Oates used the language of the Beat Generation, but retained a rhyming form in this 1961 log on USS *Alameda County* (AVB 1):

Naples is Rockin'
The Joint's alive
In berth fifty sev'n
We dig that jive.
The anchor is set
Deep in the mud,
But Daddy'O
This may cause a flood.

San Francisco was the heart of the Beat Generation, and that is where Ensign John D. Stempel found himself on January 1, 1960, aboard USS *Walton* (DE 361). He not only defied traditional poetic style, but also managed to poke a little fun at trainees (Tee Eye refers to Naval Station Treasure Island):

Tee Eye, that's where we are
Saaaan Francisco, the cool
city,
the beat city in
Cal-eye-forn-eye-aye . . .

Then in the background, the symbol grass—two-week grass
which stands for green, man, which in turn
stands for training raw material—yes, cool man,
yes, cat—green reserves

and their great green equation, cat:
BOY SCOUT CRUISES + ROUGH SEA = REGURGITATION = FRESH WATER
WASHDOWN
man, they were nautical boys, man.

And the beat goes on . . .

*Bathroom Graffiti.* Although not often written about or discussed in polite company, bathroom graffiti has long been a source of clever verse. One of the most popular, although probably originating much earlier in England, was popularized in the U.S. in the time of pay toilets. It begins "Here I sit broken-hearted, paid a dime and only farted . . ." Lt. R. R. Thomas used this age-old doggerel as the framework for his midwatch log on USS *Shields* (DD 596) as he welcomed the year 1966:

Here I stand all broken hearted;
To the beach all my friends have departed.
Tied starboard side to the quaywall are we,
South of pier six close to the sea.

Lt. Thomas concluded with both traditional log content and a New Year wish:

Commander First Fleet is SOPA this night
Keeping the watch till the sun is up bright.
When the dawn breaks all hands will shout
Happy New Year to all and that we do not doubt.

*Nursery Rhymes.* Of all the varieties of poetry that exist in the literary world and in broader circulation, it is perhaps nursery rhymes that are known to more people than any other. We learned them as youngsters, and as adults we sing them to our children and grandchildren. And they are of course known to the OODs who write the midwatch log. Here are a couple of examples.

SOC (Chief Sonar Man) E. Gembala welcomed the year 1962 to USS *Mahan* (DLG 11) with his version of the age-old *Hickory Dickory Dock* :

Hickory, Dickory, Dock.
The old year's up on the clock.
As the clock strikes twelve,

Into the new year we delve.
Hickory, Dickory, Dock.

And Ensign R. W. Bushnell, writing the log aboard USS *Nightingale* (MSCO 50), called upon his memory of *A-Tisket, A-Tasket* to usher out 1957 and welcome 1958:

A tisket a tasket
the old years in a casket
An eve without grog
We've only the log
And last years work in our basket.

- - - - - - - - - - - -

We have offered here a limited selection of the creative efforts of the young officers who brought their literary knowledge and talent to the task of writing the midwatch log. They did so in foul weather, on rough seas, amid the boredom of dry dock, in a war zone, submerged on special operations, and on lonely nights when their shipmates were ashore celebrating the New Year. The tradition of midwatch in verse is a rather remarkable idea that thousands of sailors have embraced over many years, and in this case their ability to tie their efforts to those of famous writers makes the tradition doubly interesting and impressive.

We began this chapter with a question: *To sea or not to sea?* Ensign Gregory P. Hicks addressed that question on January 1, 1974, in the midwatch log of USS *Horne* (DLG 30). In an extended entry that he called *The OOD's Soliloquy*, he used as a guide the famous speech of Shakespeare's Hamlet:

To sea or not to sea
That's a good question
Whether 'tis nobler from midnight to four
To suffer the cold and dampness of outrageous misfortune

Or to take refuge from this troubled sea
In the wardroom of the Horne
Which is portside to pier 8
And by some hot coffee end it?

Ensign Hicks maintained the Shakespearean feel of his soliloquy throughout, to this conclusion:

And thus the griping of the midwatch is
Stripped and painted o'er with the true cast of pride,
And the enterprise, of great pitch and movement,
We may never equal in dimension, but
The Horne is anyone equal in action

Although the ensign does not say it directly, we can only conclude that the inevitable answer is *Yes – To Sea!* And we can be glad these sailors answered the call—that they carried on the tradition of midwatch in verse and became poets in the most unlikely circumstances.

# SIDELIGHT 7

## A Mind of Metal and Grace

Charles J. McMahon, Jr. was born in 1933 in Philadelphia and grew up in the Germantown and East Mt. Airy neighborhoods. He attended the University of Pennsylvania on a Navy ROTC scholarship and earned a bachelor's degree in Metallurgical Engineering in 1955. He was the first of his family to attend college.

After graduation from Penn, he served aboard USS *New Jersey* (BB 62) and USS *Thuban* (AKA 19) where he held the position of communications officer and cryptographer. After serving on active duty for three years, McMahon enrolled at MIT and eventually earned a doctorate in physical metallurgy. After a postdoctoral fellowship at Penn, he joined the faculty and spent the rest of his career in the Department of Materials Science and Engineering.

McMahon was a beloved professor known for his ability to teach students how to think rather than what to think, according to many of his students. He was a pioneer in the use of digital media in education, often using video and animations in his presentations. An expert in steel fracture processes, he received many awards and was inducted into the National Academy of Engineering in 1980. His work was valuable to construction practices on bridges, buildings, and ships.

McMahon was a life-long physical fitness advocate who took up rowing in college and continued with that sport off and on into his 80s. In a tribute, his family said: "He will be remembered by family, friends, and colleagues for his lifelong commitments to charity, integrity, self-discipline, and to realizing a just and thriving society."[1] Dr. McMahon passed away in 2022 at the age of 89.

Here is his poem:

> Far out in the port of Naples
> Out across the gloomy water
> Lies the THUBAN safely nestled
> 'Gainst old Molo San Vincenzo.
> As if some great beast all fettered
> In the Mediterranean moor
> Twinlines tripled, single wire
> Hold her head out sightless staring
> At the cold and foreign shore.
> Her great eyes weeping shots of chain,
> Borth her claws upon the ground,
> Four score and ten and seventy-five
> Her fathom count in figures round.
> A hemp line doubled to a buoy
> Keeps her thoughts on home and brothers,
> Thoughts which time will ne'er be graying.
> Security watch is nightly prowling
> Against all enemies standing ready.
> Condition "B" seals her bowels,
> Waterline lights all burning steady.
> Engines lie in solemn slumber,
> Boiler spewing auxiliary steam,
> A howe to shore is sipping water,

While sailors in their hammocks dream.
In the darkened harbor ‘round us
SIXTH FLEET vessels still are lying,
COMFAIRELM their leader present,
Ships of trade their wares are plying.

So the New Year breaks upon us,
Breaks and showers hopes of peace.
Hopes that men will walk like brothers,
Hopes that war horns sound no more,
That with the sword of anger sheathed,
We’ll sail again for homeland’s shore.

# CHAPTER 7

## Phantoms of the Deep: Hauntings and Visitations at Sea

Paranormal activity reported by mariners is as old as sailing itself. Shipboard paranormal events often took the form of ghostly apparitions appearing on deck or in the bowels of the ship. These specters ranged from benevolent spirits festooned with a friendly smile merely standing watch, to entities who warned of impending doom, allowing sailors to avoid a catastrophic event. Others emit a sense of malevolence leading sailors to fear that they may be dragged by the apparition into the water's depths to an eternal gravesite. Entire ghost ships became objects of legend. Famous for several hundred years was the Flying Dutchman with its spectral crew doomed to sail the oceans for eternity. Legend suggests that looking upon it leads to potential disaster.

Why would sailors possess such an array of paranormal beliefs? Historically, sailing oceans and seas was a dangerous occupation rife with unpredictability. Humans possess an insatiable desire to make sense of their experiences. Under conditions of uncertainty and danger related to weather and ocean conditions, sailors likely sought ways to make their experiences more predictable. Conjuring spectral images that sailors believed to be related to potentially negative or positive outcomes fulfilled their need for predictability. Sailors' life on the high seas also tended to be monotonous and lonely. With nothing to see but the horizon, they likely experienced a phenomenon common to people in sensory isolation, tricks of the mind such as seeing something that is not actually there or misinterpreting an actual stimulus. A small puff of smoke or steam may appear to take on a ghostly human form.

Sailors writing deck log poems aboard modern vessels probably do not experience or report specters such as those that appeared during the vast history of sailing.[1] They do, however, pay homage to their maritime roots by writing verse that is reminiscent of spectral visitations of yore. And what better time? The midwatch on New Year's Eve is lonely duty. On a night when most people are ringing in the new year, sailors with midwatch duty must maintain vigilance under conditions of boredom, sometimes within earshot of New Year's revelers. What better way to pass the time and fulfill their duties as OODs than by conjuring up images of spectral visitors to the ship?

The most common manifestation of ship visitations in Cold War deck logs reaches back to Greek and Roman eras for inspiration. Two characters, a "New Year's Baby" and "Father Time," became the prosaic apparitions appearing in midwatch poems during the 1950s through the 1970s. The New Year's Baby may have represented Dionysus, the Greek god of fertility, festivity, and new beginnings. Father Time may have evolved from the character Chronos, the Greek god of time and the passage of time. Over the centuries, these two characters developed into very specific personifications. The New Year's Baby is usually described as wearing little clothing apart from a diaper, sometimes wearing a sash inscribed with the date of the new year and occasionally wearing a top hat. Father Time is an old man, sporting a long white beard and often carrying an hourglass or scythe. The latter is usually described leaving the ship weary after the past year's duty. The former becomes his replacement with youthful exuberance to get under way and tackle the year to come.

Many logs contain the imagery of these two mythical characters, but one log spun a yarn with delightful descriptions of their appearance as well as the OOD's and his petty officer's (PO) reactions to encountering these specters. Ensign E. C. McCann aboard the oiler USS *Hassayampa* (AO 145) in 1970 writes a long narrative describing his encounter with the New Year's Baby and Father Time that spans two and a half pages in the log. Ensign McCann opened the midwatch log by reporting a few facts about the ship's condition and a warning to the reader that he was about to regale them with a true story. The story begins with the following:

> We were just doping off when below on the well deck
> Came a rustle and a shuffle and a chill down my neck
> The ladder up forward by station one creaked
> And up through the grating two beady eyes peeked
> While visions of sabotage danced in my head
> My PO's solution was etched out in lead.
> He moved his hand slowly to the flap at his side
> Withdrew the pistol and pulled back the slide
> All of which proved too much for our friend on the ladder
> Who pulled himself topside thinking best of the matter.
> The two of us stood and blinked in the night
> Not quite believing what greeted our sight.
> My PO remarked that in spite of the light
> "Just knowing that something about this guy ain't right!"

He stood clad in sandals and costume bizarre
Showing no stain of black oil or even of tar
That he wasn't from this ship by all could be seen
Our laundry's not that good and nothing stays clean
He was dressed all in white, draped in a sail
With a white beard so long that on the deck it did trail
An hourglass upright he held in his hand
Through the stem of which trickled the last grains of sand
He shuffled toward us, his back was bent double
But stood tall on the quarter deck in spite of his trouble.

Father Time goes on to recount some of the ship's experiences over the last year and finally tells McCann that his relief is arriving and it's time for him to leave.

"Well, that's about it he said with regret
I've gotta be going, the new watch is set
My relief should report aboard anytime now
If I'm not mistaken that's he on the brow"
The old man saluted turned and departed
While with unsteady steps a little kid tottered
Up the brow and stood at the head smiling
With hand to his forehead and looked quite beguiling
"May I come aboard Sir?" He asked in voice shy
Still more than bewildered, "I guess so," said I.
His diapers were hanging down by his knees
But his white hat was square and he tried hard to please.
He could have done more about the way he was dressed
And my PO to the end remained most unimpressed.

McCann finishes the log with doubts about reporting his spectral encounter.

I looked at my PO and he looked at me,
We both shrugged together and sighed audibly,
"We can't go and report this, they'll think we're both nuts,
Needing medical discharge with no ifs ands or buts."

But I thought a bit more and made my decision,
No one believes Ensigns, so in spite of derision
I'd write down my story in the ship's smooth deck log
For Admirals at BUPERS to stare at agog. [BUPERS – Bureau of Naval Personnel]
Now one thing remained to make my tale whole,
To the old 1-MC round the corner I stole,
I picked up the mic and to those done imbibing
In my best bosun's voice called out "new year arriving!"

A couple of OODs recounted visitations by their ships' namesake. General William Eaton was an American diplomat and soldier who operated in North Africa. He is best known for leading U.S. efforts to halt "payments" to piratical North African states in the late 18$^{th}$ and early 19$^{th}$ centuries. Ensign A. J. Armstrong aboard USS *Eaton* (DD 510) in 1964 chronicled a visit by General Eaton and his grey steed.

When suddenly on a neighing grey, appeared a general in a British cloak
"Now trot you down this fandangled brow," his voice, thundering, spoke.
The pair clattered down from TIDEWATER's port and stopped on the EATON deck.
"Moor no more to AO three-one, or I may break my neck.
I am General EATON, and as the custom be
I shall inspect this New Years Eve, the ship you named for me."
(I stood with my mouth agape,
As he unhorsed his eighteenth century shape).

After reviewing ship's condition, the general says that he needs to go. But, where is his steed? Ensign Armstrong describes an odd situation where Eaton's horse finds its way up to the pilot house and gets its tail caught in the captain's chair. Eaton jumps in to free his horse,

> But the "Master of the Derne" assumed control (as at Tripoli)
> Two pulls, three tugs, a jerk, and a half-tailed nag was free.
> So when you find the Captain's chair entangled in grey hair,
> Please believe, my gentle friends, that 'twas General EATON there.

Lt. W. G. Grantham also welcomes his ship's namesake (USS *Ingraham*, DD 694) in 1968. Capt. Duncan Nathaniel Ingraham served in the US Navy until the outbreak of the Civil War when he resigned his commission and joined the Confederacy. Grantham describes the appearance of a grey-clad figure.

> His silent steps on ghostly show
> Engraved his path upon pier two.
> His rebel hat was set askew;
> -His epaulets of golden hue-
> A sabre sheathed swung shining, too.
> "This ship which bears my name
> I've come this night to see the same,
> To find if she has earned the fame,
> And has the right to make the claim,
> To Captain Ingraham's proud name."
> After reviewing specific ship information, the Captain appears satisfied with the ship.
> "I came to see and now I know,
> This ship is proud, and justly so."
> Then as the wind began to blow,
> He disappeared into the snow.

Three additional visitation poems of note appear in midwatch poems. One paints a frightening portrait of a spectral Ho Chi Minh that appears in a poem aboard the minesweeper USS *Gallant* (MSO 489) in 1966. During Market Time ops in Vietnam, Lt. j.g. George H. Riley added a surprising and timely twist to the idea of ghostly visitations.

> Seas were running high and fast
> A lookout shouted then stepped back aghast.
> Close aboard to port there slowly rose a wreck
> The wind died down and within an eerie light
> No one of us can now be sure of the ghostly sight
> The only sound was rubber sandals slapping on the deck
> A bearded figure paced the gory wreck
> His clothes hung loosely from his bones
> And all around this rotten derelict
> Came hellish wretched moans.
>
> . . . . .
>
> Against the tattered sails that hung like human skin
> That railing cursing thing was Ho Chin Minh.

A second visitation imagines a midnight visit to USS *Bigelow* (DD 942) in 1959 by Chief of Naval Operations, Adm. Arleigh Burke. During WWII, Burke commanded destroyer squadrons in the Pacific, most notably Destroyer Squadron 23 ("Little Beavers"), which played a key role in naval battles against Japan. His aggressive tactics and effectiveness earned him the nickname "31-Knot Burke." Lt. j.g. R. K. Leopold recounts the visit of the renowned destroyerman by borrowing the poetic style of Clement Moore's *A Visit from St. Nicholas*.

> Then what to our wondering eyes should appear,
> But a 4 star Admiral (no, not 8 reindeer).
> The P.O. mumbled with accompanying smirk,
> "Well, old man, here comes Arleigh Burke."

More rapid than missiles 8 sideboys they came,
As I whistled and shouted and called them by name;
"Up Connors and Bowling -fall in on the double!
Shake it McClure or we'll all be in trouble.
Line up by the gangway! Creear face toward the bow!
My God here he comes! Holy Cow! Fall in now!

..........

As I snapped to attention and piped him aboard,
At 31 knots 'cross our brow he then roared.
He was dressed all in blue from his stem to his stern,
The soot from the coal piers thus caused much concern;
The stump of his pipe he held clenched in his jaw,
In his eyes a most friendly twinkle I saw.

After "inspecting" the ship, the Admiral prepares to leave

"Adieu", he said, "right home I must hustle,
And prepare for relief by Vice Admiral Russell."
And putting his hand to his cap he saluted,
Stepped on the brow, and shoreward then scooted.
This log I thus sign; there'll be trouble, I know,
Convincing the Skipper it was CNO.

The final example of a "visitation" involves the experiences of sailors aboard USS *Arneb* (AKA 56) in 1964. Lt. j.g. C. J. Morely is standing the midwatch with his petty officer on a cold night.

A sailor's tongue bears many yarns
And many have I heard,
But never did I think that I'd
Be there when one occurred

The night was black, the wind was up,
The rain was bitter cold;
The watch and I were huddled close

Together in a fold.

The messenger was shivering,
The PO's nose was red.
"These Norfolk nights are colder than
A witch's ear," he said.

A figure appears on the pier.

Among the lines collected there
I saw an ancient man
Whose face reflected years at sea
As only sea-dogs can

The old man requests to inspect the ship and his request is granted. After touring below decks, the man emerges topside.

"I thank you, lad, for showing me
The innards of your ship,
And now I'd better make my way,
For I've a long, long trip."

The old man climbed the brow to leave,
A tear was in his eye'
He paused to scan the ship again
And breathe a heavy sigh.

And then he spoke in solemn tones,
In words so crisp and clear
That I remember every phrase
As you my voice do hear:

"I doubt if you'll believe it, men,

But I am Davey Jones,
And once a year, on New Year's Eve,
I dry my watery bones."

"I come topside to see a ship
And walk the living decks:
I come to see how matters are
On vessels that aren't wrecks."

"And I have seen a goodly ship
And you have served me well,
So now I leave for one more year
To rule my watery hell."

The night was black, the wind was up,
The rain was bitter cold,
The watch and I were huddled close
Together in a fold.

# SIDELIGHT 8

## Admiral John Hoskins: The Man with a Wooden Leg and an Iron Will

Sometimes a phrase in a midwatch poem stands out and begs for clarification. An example of this appears in the January 1, 1956, log of USS *Howard D. Crow* (DE 252).

> Far across the bay is the man who holds sway with the title of COMFAIRLANT [Commander, Naval Air Force Atlantic]
> He's a warrior tried and true who's known to all of you
> With his one leg made from a plant.

Lt. j.g. W. N. Cottrell refers to Admiral John Hoskins, who reported to relieve Capt. William H. Buracker as commander of the light carrier USS *Princeton* (CVL 23) at the beginning of the Battle of Leyte Gulf in 1944. On Hoskins' second day aboard, Princeton was hit by a Japanese bomb that crippled the ship. Capt. Buracker ordered the crew to abandon ship, but Hoskins volunteered to stay behind with a skeleton crew. The fires on board ignited ordinance stores resulting in a huge explosion that killed many of the remaining crew and blew off Hoskins' right leg above the ankle. He was rescued and sent out of the combat area for medical treatment. Ever the humorist, as Hoskins was being carried on a stretcher past Capt. Buracker he queried with a smile, "Permission to leave the ship, Sir?"

This wound normally ended the careers of sailors, but Hoskins claimed that after recovery he was ready to serve and so began his quest to be reinstated to command. When Admiral "Bull" Halsey visited him, he exclaimed, "Hell, Admiral, the Navy doesn't expect a man to think with his feet. That blast didn't knock off my head."[1] He famously chose a hospital room overlooking the naval shipyard where *Princeton's* replacement was being built. After much rehabilitation after being fitted with a prosthetic foot, Hoskins was reinstated and became the first commander of USS *Princeton* (CV 37) in November of 1945.

He eventually went on to command carrier divisions and air fleets. At one point, Air Group 81 from *Princeton* had Walt Disney design a patch that portrayed "...a saber-slinging pirate with an aircraft carrier under one arm and a peg leg firing ammunition like a machine gun."[2] Hoskins retired from the Navy in 1957 and passed away on March 30, 1964. He is buried in Arlington National Cemetery.

# CHAPTER 8

## Leatherneck Bards: Marines Embrace Unusual Duty Aboard U.S. Navy Ships. [1]

### Introduction

Marine Detachments (MarDets) aboard US Navy ships have a history as long as the Corps itself.[2] By the time the Corps reached its 165th birthday as WWII loomed on the horizon, the duties of seagoing Marines were well-established. These duties included providing ceremonial details, manning secondary gun batteries, overseeing overall ship security, and brig operation to name a few. Eventually, when the US entered the nuclear age, MarDets guarded shipboard "special" (nuclear) weapons. Most seagoing Marines served aboard aircraft carriers, battleships, and cruisers but some served as Ship Combat Cargo Officers (SSCO) aboard attack cargo ships and similar vessels.[3] One of the lesser-known activities performed by MarDet officers aboard ships was the role of Officer of the Deck (OOD) while the ship was in port. Some Marine officers while serving as OODs during the midwatch on New Year's Eve took up the mantle of bard and wrote the first log entry of the year in verse. Most Marine written poems mirrored the themes that appeared in logs written by sailors, but they also put their unique mark on their log entries.

## Marines respond to the call to write the midwatch in verse

In a review of over 10,000 midwatch poems entered in US Navy ship deck logs between 1947 and 1978, 21 were written by Marine officers serving aboard ships. Marine-written midwatch poems usually resembled those written by their Navy counterparts with respect to content and form but some differences appear.

Marine officers writing the New Year's Eve midwatch recognize the tradition and, rather than issue complaints about writing it in verse, they seem to accept it as a *fait accompli*. They will do their duty and follow tradition. An example of this appears in USS *Montrose*'s January 1, 1965, log written by 1st Lt. Mike H. Roulier in an interestingly organized poem.

THE NEW YEAR'S MID WATCH

OR

AN AGONY IN THREE FITS

------FIT THE FIRST----

On a midnight clear and cold
With his slow and halting rhetoric
A poor displaced Marine was told
To carry on the tradition metric

Make the log recite in running rhyme
Information both trivial and practical
Include the event and its time

And some data autobiographical.

After a second fit wherein Roulier rhymes ship conditions and status, he closes with the third fit.

----------FIT THE THIRD---------
Now appears the new
While the old year fast is dimming
Let's all the bidding review
And never our wheels be spinning

Montrose is our name
"Will do" our answer
Our response is always the same
--"Of course we can, Sir"

With My Apologies to Poets
And the U.S. Navy
Mike H. Roulier
1st Lt. USMC

Several stanzas written by Capt. B. A. Adams, USMC, (USS *Macon,* CA 132) who received the Silver Star Medal and three Purple Heart Medals for action in Korea, exemplify his perceived call to duty as a keeper of the rhyming tradition.

Whoever thought to write in rhyme
Had naught to do to pass the time,
But tradition is tradition and here we go
With a brand new log and a ho-de-ho
(Ho and go, they rhyme you know,
Like futch and glockenspiel).

Later, Adams writes

I did it, I did it, I knew that I could;
I 'rit' this in rhyme like Longfellow would.
My work is now finished 'till dawn early light.
A Happy New Year to all and to all a good night.

But at 0220, a sailor is cited by the commanding officer for several serious offences that must be recorded in the official ship's log. Adams acknowledges this in a postscript prior to naming the insubordinate sailor and his offences in prose.

> P.S. It's criminal, it's criminal, a horrible shame
> A sailor's good name to bard or defame,
> Yet duty is duty and reports I must add
> Not written in rhyme – I'm a terrible cad!

After the brief deviation from rhyme to record the sailor's offences, Adams closes with

> If you've read all of my song, my thanks I extend
> And repeat one more time this is really the end.

In one of the stronger statements made by a Marine articulating his duty to follow tradition, 1st Lt. C. R. Keith, USMC (USS *Ticonderoga,* CVA 14) writes

> The OOD has taken the deck
> As is typical of a good leatherneck
> And in standard Marine Corps tradition
> The OOD has vowed to complete his mission

**Marine references to lack of alcohol while on watch**

As with their Navy counterparts, Marines writing midwatch poems frequently mention alcohol or the lack thereof. 1st Lt. W. M. Carlson, USMC (USS *Thuban*, AKA 19) paints a celebratory picture, but only for those ashore. Carlson even spells the name of his ship incorrectly in order to maintain the rhyme.

> Ear splitting noise tells the New Year is born
> Steam to all whistles and lips to the horn
> Sirens a'screaming and bells are a'peeling

'Twould cause a sober man to start reeling
'Most the team's on the town sucking up bourbon
This the crew of the USS Thurban
Phone service and water emit from the pier
For the latter, no doubt the crew'd prefer beer.

While aboard USS *Washburn* (AKA 108), 1st Lt. A. D. Brewin relates his observations about his own crew as well as an adjacent moored ship USS *George Clymer* (APA 27).

This eve our crew is downing brew
While snug in port we lie;
NAVSTATION (Dago), 4 the pier
Our berth is forty two
Old George's crew came tramping through
At all hours day and night
We know their home when aboard they roam
By the delicate scent of brew

**Marines make sure that it is clear who is on duty**

After their midwatch entry, OODs must apply their signature to make the log official. Signatures are usually followed by the writer's rank and service branch. "USMC" after the signatures makes it clear that the writer is a Marine, yet fully half of the 20 Marine-written logs also contained a statement that ensures that there is no mistake about who is on duty. A. N. Manella, 1st Lt., (USS *Hancock*, CVA 19) begins his midwatch entry with

It's an honor I'm told and a sight to behold
Amid frolicking, bells, gongs, and screams
The first watch of the year – be assumed and it's clear
By a Lieutenant of the U.S. Marines.

First Lieutenant G. R. Johnston, while on USS *Mathews* (AKA 96) as Ship Combat Cargo Officer, finishes his midwatch poem with

> And throughout the midwatch, this big gray machine
> Was guarded by her only U. S. MARINE.

M. M. Bruner, 1st Lt., concludes with

> The Exec's on board, the Captain's not on the scene
> But no one is worried, the OOD is a United States Marine.

A. D. Brewin, 1st Lt., implies divine intervention in his role aboard the transport *Washburn* (AKA 108).

> Our Navy dear with many a tear
> Provides for all our wants;
> But God indeed fills every need
> When our Marine came here!

J. A. Shepherd, 2nd Lt, USS *Coral Sea* (CVA 43) states,

> About the ship there is not a sailor to be seen
> The OOD is a United States Marine.

During the midwatch, Lt. Shepherd learned of a crew member returning to ship after being absent without leave. He concludes the poem with

> I learned at 2355 [redacted name and service number] returned from over the hill
> And I in disgust spilled my ink and returned to my quill.

With a nod to the quality of the crew and with a little swagger on the part of the writer, the following appears in the midwatch log of USS *Midway* (CVA 41):

> But one resolution I would like to make
> Is that our proud crew remain so great

And as they sleep after a night of cheer,
That they sleep sound and have no fear,
Because they're guarded as they rest
By the Midway Marines, the very best.
C. S. Sharples, Jr., Capt., USMC

First Lieutenant G. G. Evans, Jr., of USS *Ranger* (CVA 61) finishes his midwatch poem by acknowledging his duty to write a poem and recording his service identity.

Naval tradition dictates that this be done in theme;
At long last it's finished by a U. S. Marine

Finally, C. R. Keith defies grammar to close out his midwatch poem with a phrase that appears in several other deck logs suggesting that there was some borrowing of wording over the years.

Well, it's over, it's done – Navigator, please note
If the log doesn't rhyme, at least it was wrote.

The same final line appears in the log of USS *Fechteler* (DD 870) written by Lt. R. O. Pyle, in 1954 and was published in *All Hands* in December of that year. Lt. j.g. C. E. Pausa also uses the closing phrase in his deck log poem aboard USS *Montrose* (APA 212) in 1956.

## JFK Memorial Poem

The pride in being part of the Marine Corps shown in the poems above is part of a powerful loyalty to the Corps. Part of that loyalty is to the chain of command, and at the top of the chain sits the Commander in Chief. On June 6, 1963, President John F. Kennedy visited USS *Kitty Hawk* (CVA 63) during a fleet review. He toured the ship, watched operational demonstrations, and made remarks about the importance for the U.S. to control the seas. The President must have made a lasting impression on one of the young Marine officers. Capt. J. J. Burke, Jr., CO of the Kitty Hawk MarDet wrote a tribute to JFK in the 1964 New Year's Day midwatch log. Of the thousands of logs reviewed, this is the only one that memorialized a leader with such eloquence and poignancy.

00-04 In Memoriam

It is truly fitting that we of this mighty aircraft carrier pay our last farewell to our departed shipmate.
He was a young man, dynamic in mind and personality who was
destined to be the great leader of our nation.
This was the world leader who on January 20th in the year of our Lord
Nineteen Hundred and Sixty-one, observed not a victory
of party but a celebration of freedom.
Yes, this was our shipmate.

This was the gentleman who paid honor and praise to us on that cool
day in June in the year of our Lord Nineteen Hundred and Sixty-Three
This was the same man who walked about our decks exhorting us in
the performance of our tasks.
Likewise, this was an individual who braved the elements of the
sea in a world war not so long ago.
This was our shipmate.

Forty-two times our hearts pounded at the roar of the guns but only half rang out in joy…..
the other half cried out, "Farewell Shipmate."

Dazed and bewildered our minds could not comprehend, "How could
two cursed shots rob a great nation?"
And our hearts continued to cry out, "Farewell Shipmate."

Bear with me now as I must honor a custom and tradition that he,
I am sure, once honored some years ago:

The poet adds several lines regarding the ship's condition here before finishing his tribute.

And now that tribute has been paid to custom and tradition,
may all my shipmates join with me in his famous words,
"And so, my fellow Americans, ask not what America will do for you, but what you can do for your country.
My fellow citizens of the world, ask not what America will do for you, but what together we can do for the Freedom of Man."
Brother of the sea, your words shall always be treasured
and we shall always remain poised and ready to continue
to carry the flag of freedom.
May your soul rest in peace and may God bless all the
shipmates who have gone before us and also those who
shall come after us.

J. J BURKE, Jr.
CAPTAIN
"SOLDIER OF THE SEA"

# SIDELIGHT 9

## Women in the Navy[1]

Women began unofficially serving in the US military as far back as the Revolutionary War. They mostly served as caregivers of ill and wounded soldiers and sailors. It was not until the American Civil War that the US Navy began to actively recruit women to serve as nurses on the first hospital ship, USS *Red Rover*, leading to the first documented official service of women in the Navy. A permanent nursing corps emerged in 1908 when Congress established the Navy Nurse Corps and selected 20 initial members often referred to as "The Sacred Twenty."

After the outbreak of WWI, the US Naval Reserve Act of 1916 established the rate of Yeoman for women (Sometimes they were referred to as Yeomanettes). These women mainly served in clerical capacities, freeing up men for more combat-related duties. Fast forward to WWII, President Roosevelt signed into law an act that established Women Accepted for Volunteer Emergency Service (WAVES) that allowed women to serve in both officer and enlisted ranks and rates paving the way for future advancement.

While women in the Navy made some strides forward in the years after WWII, it was not until the 1970s that the pace of advancement of women quickened. This decade saw the appointment of the first woman admiral, the first woman naval aviator, and the first admission of women to the US Naval Academy.

Women continued their progress into the 1990s and into the 2000s by being assigned to ships and eventually to submarines. In a few cases, women took command of combat vessels including a frigate and an aircraft carrier. In 2023, Admiral Lisa Franchetti reached the pinnacle of Navy service by being appointed Chief of Naval Operations (CNO) placing her on the Joint Chiefs of Staff.[2]

Out of the over 10,000 deck log poems reviewed, we found only one that was written by a woman. Lt. j.g. Patricia Summers, a nurse aboard the hospital ship USS *Repose* (AH 16), wrote the following:

> At the birth of 1970
> As the last year quickly died
> The ship stood moored to Alava pier
> By her starboard side.
> At birth two, U.S. Naval Base
> Republic of the Philippines
> The proudest ship in this Navy and fleet
> Stood spit and polish, all brite and clean
> The standard nylon mooring lines
> Doubled fore and aft
> Securely held the stalking form
> Of this seaworthy craft.
> Our stalwart leader, Robert F.
> Today assumed the role
> Of Senior Officer Present Afloat.
> For whom the bell will toll.
> Close by us are the *Tripoli*,

The *Hassayampa* aft,
Ships of the U.S. Pacific Fleet,
Yard and district craft.
The engineering plant, I hear
Is in a state of naught.
Cold Iron Status has given them
Relief so seldom brought.
The ship sits here, receiving from
The pier their ten-fold service;
It's good I won't be here for long –
To watch them makes me nervous!
The Captain is ashore tonight
Likewise much of the crew
I wish you all a pleasant stay,
And a Happy New Year, too.

# CHAPTER 9

## Midwatch Mischief: Navigating Unusual Rhymes in Deck Log Poetry

As outlined in Sidelight 1, OODs faced with midwatch duty on New Year's Day often complain about the pressure to write the log in verse. Few sailors write poetry and even fewer possess the proficiency to write good poetry. The expectation to include nautical jargon in the log makes the task even more challenging. One expectation involves reporting the ship's condition of readiness that defines the ship's current condition affecting its ability to react to attacks or emergencies in effective, often life-saving ways. One of these conditions, YOKE, appears frequently in logs. Most often, OODs end up rhyming this condition with "joke." "YOKE" is a condition sometimes used at sea when waters are calm during peacetime operations.

While an almost endless list of naval jargon may be included in midwatch poems, sailors have wide latitude to deviate from "just the facts" and exercise their creativity in their writing. Some sailors avoid the rhyming issue altogether by writing in free verse, but those poems are rare. So, writers steel themselves for the task at hand and attempt to rhyme "by hook or by crook" using an array of methods.

### Rhyming on Repeat

Instead of using a traditional series of rhymes, some sailors simply used the same ending on multiple words, creating an imitation rhyme. Lt. Elton E. Hankins aboard USS *Bashaw* (AGSS 241), 1964, wrote

> Here we are at the U.S. Naval Submarine Base, Pearl Harbor, Hawaii,
> The navigator whispered silently.
> And moored starboard side to berth S-13,
> The Executive Officer stated fixedly.
> With miscellaneous services coming from the pier,
> The Engineer volunteered receptively.
> The crew's been mustered and all hands are present,
> The Duty Chief reported absently.
> Material condition BRAVO is set,
> The Gunnery Officer shot out disarmingly.
> Various yard, District and Pacific units are present,
> The Operations Officer remarked fleetingly.
> I am Senior Officer Present Afloat,
> CINCPACFLT smiled up buoyantly.
> And that ends this chore for another year,
> The OOD sighed logily.

Lt j.g. R. E. Ryan must have been dubbed the official bard of destroyer leader USS *Mitscher* (DL 2). He wrote the midwatch poem in both 1956 and 1957. In the latter year, Ryan opened his log with

> It's New Year's and I must be philosophical
> And try to write a log that is shakespearical,
> Although it's hard to be all sweet and nice and lyrical,

About a ship whose movements are so quizzical

His poem that takes up an entire typewritten page in the log continued with many clever rhymes in decent meter and he returns to his opening scheme to close out his log entry.

And so my task is done and I'm not whimsical,
The night is long and the cold has got quite physical,
So I close with a prayer that is old and metaphysical:
"Even though our bugler's not the same as Satchmo,
At least, Lord, give us one more crack at GTMO."
[GTMO Guantanamo Bay]

Aboard the oiler USS *Salamonie* (AO 21), Ensign E. J. Bride concluded his midwatch log with a plea to the Navy.

(Just one final thought for the Bureau to ponder
Of Navy life our lonesome girls'd be fonder
If husbands and boyfriends weren't sent "over yonder").

Finally, Ensign M. R. Charley aboard USS *Zelima* (AF 49) drew inspiration from Poe's "The Raven" when he wrote

I WAS NAPPING, WHEN A TAPPING ON THE OVERLAPPING
COPING WOKE ME. GAPING, YAPPING, GROPING, I WENT
HOPPING, LEAPING HOPING THAT THE TAPPING ON THE
COPING WAS JUST THE CAT AND MERELY THAT. BUT
QUOTH THE MESSENGER:
"00 – 04" AND NOTHING MORE.
ONCE UPON A MIDWATCH DREARY, EERIE, SCARY, I WAS
WEARY. I WAS WARY, FULL OF WORRY O'ER AF 49. THIS

FEELING INSALUBRIOUS SOON MADE ME MOST LUGUBRIOUS
SO ATTENTIVELY I LISTENED TO THE SKINNY BEING
PASSED. ............

**The Watch Officer's Wordsmithery: Unusual Words Found in Deck Logs**

During the midwatch OODs have plenty of time to conjure up rhyming words for their poems. They may be assisted in their search for rhymes by their fellow sailors on watch. It's easy to imagine an OOD asking fellow sailors, "What's a word that rhymes with ____?" Sometimes the OOD and/or crew come up with rhyming words that are unusual to find in deck logs. In some cases it is difficult to imagine that any other deck log in the past has ever used that uncommon word. Here are some of those examples from a wide range of ships.

Moored portside in Charleston,
At MINELANT Pier Sierra,
Our glowing decorations seem,
A holiday tiara

Ensign J. P. Harris
USS *Affrey* (MSO 511), 1970

To write this was a poor decision;
On the feast of the Lord's Circumcision.

Ensign R. P. Crane
USS *Truckee* (AO 147), 1958

From the pier we receive miscellaneous services;
While steam wisps from boiler #1 swirl like dervishes.

Ensign J. E. Tope
USS *Columbus* (CA 74), 1959

The wintery breeze blows from the shore,
Bikinis are gone from Cote D'Azure

Ensign Phillip D. Lank
USS *Conyngham* (DDG 17), 1965

Around us lie ships, from the vast Pacific,
The AEOLUS, some tugs, and others magnifique.

Ensign P. J. Tiernan
USS *Bexar* (APA 237), 1959

The snipes have been playing boiler roulette,
["snipes" - enlisted personnel working in the engine room]
With ables and bakers, an absurd minuette.

Lt. j.g. L. D. Trabert
USS *Saratoga* (CVA 60), 1969

Writing doggerel verse with feet trochate
To greet the New Year as old salts dictate.

Lt. j.g. W. R. Brandon
USS *Constellation* (CVA 64), 1971

Finally, while standing in harbor it is common to list the other ships present. Large anchorages may host many ships making their inclusion problematic for a poet. One OOD solved this problem with the following:

> Many ships are here in the vicinity
> A complete list would be asininity.
>
> Lt. j.g. B. M. Bell
> USS *Esteem* (MSO 438), 1956

**Not Quite Common: Unusual Rhyming Words**

Some OODs did not venture quite as far into the realm of the extraordinary to achieve their rhymes. Aboard the Acme-class minesweeper USS *Advance* (MSO 510), Ensign B. W. Thompson opened the New Year's Day midwatch log with

> 'Tis New Year's Morn, and verily,
> Through night that's dark and calm,
> We're steaming solitarily
> Off the coast of Vietnam.

At the time, *Advance* was participating in Market Time, a multi-year operation designed to interdict supplies and munitions flowing by sea from North to South Vietnam.

References to the New Year's Eve tradition of alcohol consumption appear often in midwatch poetry. Alcohol is prohibited aboard ship (see chapter 2), but while moored in harbors sailors on liberty take the opportunity to imbibe. OODs chose interesting ways to document the alcohol consumption by their fellow sailors.

Though we be a ship most modern and steel-skinned
The sailors are returning with three sheets to the wind.

Lt. J. L. Szczecinski
USS *Canberra* (CAG 2), 1966

Throughout the ship is set condition YOKE MODIFIED
And pardon our saying, but half the world seems pie-eyed.

Ensign J. E. Tope
USS *Columbus* (CA 74), 1959

Though Okinawa's not far in fact, land is so close it's enticing
Buckner may not be great but at least it's a place where the main brace we all could be splicing [Splicing the Mainbrace is a Navy term for drinking alcohol.]

Lt. T. W. Erikson
USS *Pomfret* (SS 391), 1962

It was not just alcohol consumption that led to rhymes using unusual words and phrases as evidenced by the following excerpts:

But the night was clear and the number 8 was running
To bring back all of those who had been on the town funning.

Lt. j.g. R. G. Wells
USS *Oriskany* (CVA 34), 1962

To return to the business of logging the facts
With poetic but accurate purpose,
Our water and steam, our coffee and cream –
The dock provides all – except burp us.

Lt. C. E. Gurney
USS *Farragut* (DLG 6), 1961

A year ago in Newport we were
Snow and cold and ice and brrrr;

Lt. E. C. Coffey
USS R. E. *Kraus* (DD 849}, 1969

Lt. j.g. J. A. Sears, Jr. aboard USS *Mullinix* (DD 944) in 1959, wrote several rhymes with unlikely word origins including a curious reference to one of the leading brokerages of the time.

Port side to pier three using standard manila
My coat lined alpaca, or perhaps it's chinchilla
Cold iron prevails in our hull quite impervious
With the shipyard providing the needed pier servious

Referring to the condition of the ship being monitored by security, Sears, Jr. wrote:

For it's checked by security watches terrific,
Merrill-Lynch, Pierce, Fenner, and Smithic.

He finished with the following salutation:

Anapestic quadrameter couplets, Adieu!
To my whole duty section a Happy Year New.

Finally, Lt. J. S. Sheets aboard the submarine USS *Charr* (SS 328) in 1959 summed up the bane of any large organization like the Navy: Generating reports.

Starting the New Year is easy to shirk.
But ending the old – takes paperwork.

**Rhyme Before Reason: The Art of Invented Words**

When OODs' muse fails them and threatens to scuttle their rhyme-craft, they sometimes resort to creating words that do not appear in the dictionary. After all, poetry should rhyme!

Ensign L. G. McIntire closed out his 1962 poem aboard USS *Warbler* (MSC 206) a Bluebird-class mine sweeper.

And so mid log, this ends the requirements
For poetry used throughout your entirement
I now shall partake of a little desirement
Of laying down and enjoying retirement

L. G. McIntire, Ensign, USNR

Lt. j.g. David O. Weber, who became a writer of note after his naval service, plied his considerable ability at wordplay to a midwatch log while aboard USS *Higbee* (DD 906) in 1963. After providing typical information about the ship's condition, he catalogued the ship's journeys over the previous year.

1 JAN
Finds us in Japan
Feasting and wining and wenching and otherwise getting
out of shape and becoming homeportly.

*Higbee* visited Burma in 1962.

Then through the Straits to the Golden Pagodas of Burma,

Up the shallow and tortuous river with scarcely a murmur
Till rounding a bend, comes the first view of shimmering Rangoon
All lush and verdant - - teaked, camphored, palmed and bamboon.

-------------

It was hard to find a saloon
In Rangoon.

It was then on to Japan.

And who could forget The Black Ship Festival of Japanese-American friendship in tiny Shimoda?
What more friendly and hospitable place, and what more worthy occasion, could one possibly hope for to go ta?

After arriving at Hawaii

We have a peeky
At Waikiki

Aboard the escort carrier USS *Windham Bay* (CVU 92) in 1957, Lt. K. N. Miles wrote

The O.D. said, as he nodded his head,
To the Bos'n standing near,
"Come out of the fog, and we'll write this log
In iambic pentamateer."

--------------------

On December five six, two seven one nine two six
And with Zulu attached dates communication
When COMISTUS SERV PAC, tossed us Santa's sack
He said, "Depart with fee-licitation."

Some OODs followed suit in the quest to rhyme at all costs. Lt .j.g. J. R. Lewis III, while aboard USS *Alacrity* (MSO 520) in 1969, wrote.

> Come all ye young Ensigns with bright perspicacity
> And a JG will rhyme ye New Year's Midwatch aboard *Alacrity*
> In childish meter and rhythm quite doggerel
> To leave your young brains in all fuzzy fuggerel.

Other examples of creating tortured rhymes include:

> At US Naval Station, Mayport, Florida.
> The weather's not fair; it's downright horrida.
>
> Lt. j.g. J. D. Lange
> USS *Ault* (DD 698), 1964

> Moored as always before we are,
> Keeping station on Mr. Fuji bar.
> It's a taut watch we're standing with great vigah;
> Half our watch wish they had just one jigah.
>
> Lt. j.g. K. E. Farnsworth
> USS *Jupiter* (AVS 8), 1964

> The pier is D, the berth is five
> and the services are numerous
> The number of HOOD is six five five
> and the duty section gloomerous.
>
> Lt. Edward H. Monroe, Jr.
> USS *Remey* (DD 688), 1961

Lt. J. W. Bird, Jr., in 1958 serving on the submarine USS *Picuda,* did not appear to be thrilled to have the New Year's Eve duty when he wrote:

> Here at the U.S. Naval Station, Key West, state of Florida
> The duty officer composes verse in a state of morbida.

But apparently, one sailor livened up the evening.

> In control the below decks watch is singing "All's Well"
> On the bridge topside is tooting the whistle and ringing the bell
> The rest of the section is on deck wearing dungarees
> One is shooting the very pistol, the others are shouting whoopsies.

The "very pistol," or Verey pistol, is a snub-nosed, single-shot, breach loading pistol designed to fire flares. Doing so for celebratory reasons in port, while not unheard of in maritime history, was not likely permitted based upon Lt. Bird's characterization of "whoopsies."

Finally, Ensign W. C. Montgomery (USS *Thuban* – AKA 19), like many of his fellow sailors who have New Year's Eve duty, described his night compared to those who were on liberty.

> For on the beach while men make merry
> We sit in Norfolk Naval Shipyard on our derrierie.

## Rhyming Proper Names: What Rhymes with Orange?

One of the most difficult challenges of midwatch poets involves the frustrations of rhyming names of people, places, ship names, and hull numbers. The rules for deck log content dictate that at the very least, the log should include where the ship is located, what ships are nearby, who is the overall commander of that group of ships, and the ship's operational condition. Navy jargon is rife with ostensibly un-rhymable acronyms such as SOPA (Senior Officer Present Afloat), DEPCOMSUBLANT (Deputy Commander of Submarines, Atlantic Fleet), and COMPACFLT (Commander, U.S. Pacific Fleet) seemingly designed to thwart the midwatch bards' ability to create a rhyme scheme. One solution would be not to put such acronyms or proper names at the end of a line. Yet, sailors sometimes paint themselves in a proverbial rhyming corner by doing so, requiring them to work hard to find a decent rhyme.

Lt. S. W. Sigmund did a good job of rhyming a variety of proper names and hull numbers in his 1968 midwatch log from USS *Catfish* (SS 339). He also did not shy away from rhyming Navy acronyms.

> We should all be ashore this first of the year
> Enjoying the songs, gin, whiskey, and beer,
> But we sit here with duty, alas and alack,
> In the next of four subs at NAVSUBSUPFAC [Naval Submarine Supplies Facility]

Lt. j.g. R. M. Bowen, in 1961 aboard USS *Lake Champlain* (CVS 39), used a French phrase to rhyme a frequently appearing acronym.

> Our COMCARDIV 20 of course is the SOPA
> And he makes darn sure that we don't make a faux pas.

Lt. R. A. Ruble focused on several of the issues addressed in this chapter while writing the log for USS *Dogfish* (SS 350) in 1962.

> SOPA is DEPCOMSUBLANT in
> DIABLO (SS 479)
> If we can say that without pantin'
> Our fitness program is doing fine.

*Rhyming Ship Names and Hull Numbers.* To the casual observer, ship naming conventions, designations, and hull numbers may be confusing. There is rhyme and reason for how the Secretary of the Navy assigns these descriptors. What does "USS *Catfish* (SS 339)" mean? USS stands for United States Ship. *Catfish* is the vessel's name. The SS is the designation for submarine. The 339 hull number roughly designates where in the line of production *Catfish's* type of vessel occurred. Thus, 339 would be older than 340. From the 1920's through the 1960s submarines were named after sea creatures, mostly fish. Lt. Sigmund's 1968 poem cited earlier tackles this issue of rhyming hull numbers.

> The numbers we carry so large on our hulls
> Cause the poets much anguish and scratching of skulls,
> As the rhyming of numbers can be quite confounding –
> Toss in some letters – your problem's compounding.
>
> The ROCK is the old boat – can't find many more
> With a number as low as Two Seven Four.
> Though REDFISH looks old, and barely alive,
> She sports the high number of Three Ninety five.
>
> In contrast, the DIODON, nearly modern of line,
> Wears a quite-in-the-middle Three Forty Nine.
> Now that, after all, wasn't really so bad!
> Rhyming numbers with words, though taxed me a tad.

Other examples of the struggle to rhyme ship names include

> Mother MARKAB nests the ALUDRA and alongside her the KOINER,
> While PICKAWAY and BELLATRIX are very pleased to join 'er.
>
> Lt. j.g. D. W. McCarty
> USS *Interdictor* (AGR 13), 1964

This is the United States Ship FRANCIS MARION,
Two and a half years busy troop carryin'.

Ensign R. S. Speer
USS *Francis Marion* (APA 249), 1964

It's not just our men this weather is frostin',
The FORRESTAL's here and so is the BOSTON.

Lt. j.g. T. D. Jaroch
USS *Allen M. Sumner* (DD 692), 1966

The ship tied to starboard is the destroyer O'BRIAN
Hull seven two five of this no denyin.

Ensign R. S. Corsette
USS *Maddox* (DD 731), 1964

With our phone we may invite
COMFAIR and COMDESLANT
In the middle of the night,
For we're assured by the light,
They're at Quonset and in YO-SE-MITE.

Lt. J. P. Jurich
USS *Charles R. Ware* (DD 865), 1961

*Places: Nothing Rhymes with Norfolk.* Names of cities and bodies of water provide challenges to poets, but there are ways to handle them.

We're in the waters of old Hong Kong
It has now been three days, not a thing has gone wrong.

Here in the port a British Crown Colony
Our position could be fixed no better by Ptolemy.

Ensign D. A. Wright
USS *Firm* (MSO 444), 1959

While moored in Genoa on a chilly morning:

For in "sunny" Itterly
The wind howls bitterly.

Ensign J. M. S. Finney
USS *Cassin Young* (DD 793), 1957

As PHIBRON TEN flagship we toured the Caribbean
In voyages more purposeful than a mere cruise Kon-
Tikian.

Lt. j.g. R. A. Olson
USS *Boxer* (LPH 4), 1960

Lt. j.g Michael Fanin aboard USS *Hammerberg* (DE 1015) in 1966 surely wins the prize for successfully navigating the problems surrounding rhymes of proper names.

Happy New Year to our leaders in Newport,
And to our colleagues holding the Fort
Down at Cape Cod,
Toast sixty-six and eat a scrod
For us in Beantown.

---------

Other ships, port and starboard, fore and aft,
Are LANTFLT types and harbor craft,
And to them we say, "All's well in hand."
So you people doing the town strike up the band

..........................

And COMCRUDESLANT. A chorus of "Home on the Range"
Maestro , for the Navy motion picture exchange,
Some pickled herring on crackers Ritzy
For Navy Secretary Nitze
We wish straight fairways for Captain Lukas,
Whose third stripe was not a fluke, 'cause
We've seen him in action.
Greetings made, our poem within a fraction
Of being complete, we'll only ask ya
To join in a thought for tomorrow: Go Bama, Beat Nebraska.

**Harbor Smells: Ode to Eau de Tuna**

US Navy ships sometimes dock in harbors along with merchant and fishing vessels leading to rare references to the odors experienced while on watch.

The Atlantic Fleet is well represented
Together with fishing craft (most powerfully scented).

Ensign S. P. McDonnell
USS *Kirkpatrick* (DER 318), 1958

Chief Machinist's Mate J. H. O'Neal cleverly summed up the position of USS *Henrico* (APA 45) and the odor he experienced by invoking a well-known trademark.

Bethlehem is where we are,
And sure enough we have a star,
It doesn't shine, but smells instead,
The "Star Kist" scent of tuna dead

Finally, misspellings in deck logs occur from time to time. But Ensign T. F. Bridges ends his log aboard the stores ship USS *Rigel* (AF 58) with an unusual misspelling.

> This completes my entry in the log book tonight
> You'll find no mistakes; each word is spelled wright.

# SIDELIGHT 10

## The Lone Haiku

We were surprised to find only one out of the 10,000+ poems that we reviewed written as haiku. A Japanese poetic form, the haiku consists of three lines with seventeen syllables in a 5-7-5 pattern. Ensign Yasuto Tana penned his log as a series of haiku while aboard USS *Blue* (DD 744) in 1966.

Yasuto Tana's father, Daisho, came to the US in 1928 as a Buddhist missionary and his mother, Tamoe, arrived in 1938. Tamoe was a renowned poet of the "tanka" form, similar to the haiku but with seven lines, featuring the syllable count 5-7-5-7-7. At the outbreak of WWII, the Tana family experienced the ignominy of being interned along with thousands of others of Japanese descent. The FBI arrested Daisho as a possible enemy alien because of his association with Japanese language schools that were thought to be propagandists for the Japanese government. He remained in custody separated from the rest of his family until after the war. Tamoe and her two sons Yasuto and Shibun were sent to the Gila River Detention Center in Arizona. Tana's third son, Chinin, was born during their internment. The family reunited in 1946 and their fourth son, Akira, was born in 1952.

Yasuto grew up in Palo Alto, CA, and graduated from Cubberly High School in 1958. He excelled in football and track and was his senior class president. He attended University of California, Berkeley and earned a bachelor's degree in electrical engineering in 1964. While enrolled at Berkeley, Yasuto continued his athletic activities on both the track and football teams. He earned his commission as an Ensign in the US Navy in August of 1964 and eventually retired from the US Naval Reserve at the rank of Lieutenant Commander. Tana died on December 24, 2022, at the age of 82.

Here is his complete midwatch deck log, a series of six haiku:

00-04 New Year's Haiku

A dark night precedes
The New Year's quiet arrival
In Long Beach Harbor.

Silent calm waters
Encircle a nest of four
At Berth Fifteen Three.

Like snakes, pier cables
Swarm out over the cold nest
Providing power.

For command power
Mine Forces of Pacific
Is Senior Afloat.

In holiday dress
Destroyer Blue, Seven Four Four
Emits brilliant light.

Light cast by the ship –

Symbol of birth at the end
Of a dying year.

Y Tana
ENS USNR

# CHAPTER 10

## Midwatch Deck Log Poetry Aboard US Submarines: The Early Cold War Years [1]

Poetry about submarines and by submariners has a long tradition. A poem appearing in the April 4, 1925, issue of *The Dolphin* (SUBASE New London Newsletter)[2] provides an excellent example and surely must have been written by a submariner of the time.

> Born in the shops of the Devil,
> Designed in the brains of a fiend;
> Filled with acid and crude oil,
> And christened "A Submarine".
>
> The poets send in their ditties,
> Of Battleships spick and clean;
> But never a word in their columns,
> Do you see of a submarine.
>
> I'll try and depict our story,
> In a very laconic way;
> Please have patience to listen,
> Until I have finished my say.

We eat where're we can find it,
And sleep hanging up on the hooks;
Conditions under which we're existing,
Are never published in books.

Life on these boats is obnoxious,
And that is using mild terms;
We are never bothered by sickness,
There isn't any room for germs.

We are never troubled with varmints,
There are things even a cockroach can't stand.
And any self-respecting rodent,
Quick as possible beats it for land.

And that little one dollar per dive,
We receive to submerge out of sight;
Is often earned more than double,
By charging batteries at night.

And that extra compensation,
We receive on boats like these;
We never really get at all,
It's spent on soap and dungarees.

Machinists get soaked in fuel oil,
Electricians in H2SO4;
Gunnersmates with 600W,
And torpedo slush galore.

When we come into the Navy Yard,
We are looked upon with disgrace;
And they make out some new regulations,
To fit our particular case.

Now all you Battleship sailors,
When you are feelin' disgruntled and mean;
Just pack your bag and hammock,
And go to "A Submarine"

Sometimes early submarine poems were figurative rather than literal, such as one titled Submarine, published in 1917 by playwright Eugene O'Neill, who sailed with the Merchant Marine for three years. The poem is a metaphorical commentary on the materialism of society.

> MY soul is a submarine.
> My aspirations are torpedoes.
> I will hide unseen
> Beneath the surface of life
> Watching for ships,
> Dull, heavy-laden merchant ships,
> Rust-eaten, grimy galleons of commerce
> Wallowing with obese assurance,
> Too sluggish to fear or wonder,
> Mocked by the laughter of waves
> And the spit of disdainful spray.
>
> I will destroy them
> Because the sea is beautiful.
>
> That is why I lurk
> Menacingly
> In green depths.[3]

Like their counterparts aboard surface vessels, submariners became maritime bards during the midwatch on New Year's Day and wrote deck logs in verse. Submariners have always considered themselves a different breed, so the question arises—are poems by these submariners a different breed of poetry compared to poems found in surface vessel logs?

### How are submarine midwatch poems like those of surface vessels?

*Complaints about having the mid on NYE and having to write in verse.* Regardless of whether a vessel is a submarine or surface ship, logs are rife with gripes about pulling duty on New Year's Eve. These complaints tend to focus on the pressure to produce a poem, the lack of alcohol on the ship, and the absence of female companionship, at times combining all of these complaints in the same stanza. By far, the pressure to write a log in verse is the most common complaint.

LT. j.g. H. D. Barker aboard USS *Pomfret* (SS 391) in 1956 finishes his midwatch log with comments about the quality of his poem and his frustration of a New Year's Eve sans alcohol.

As we sit on board with fun out of reach,
And the liberty party on the beach
Is hot bathing, partying, and guzzling beer,
The best I can do is wish Happy New Year.
I am going to quit writing this miserable "pome",
Our New Year will start upon Pomfret's return home.

Men of the sea often rely upon the intervention of maritime deities in times of trouble. According to Lt. Cmdr. R. E. Burns (USS *Wahoo* (SS 565), 1969) midwatch log-writing is one of those occasions.

There's a legend told by Seamen,
That when logs are written free and
In a verse on New Years Day;
That the soul of Neptune Rex
With a special sort of hex,
Tells the writer what to say.

Well, I'm banking on that story
To help me in my tale of glory
On this very special day;
And I hope the good king's spell

Will make words flow swift and well
And in a most poetic way.

Midwatch writers have also frequently invoked the names of poetic geniuses of yore. Possibly to lessen the fear of being viewed as a poetic lout, midwatch poets often contrast their own efforts with the skills of the masters. From the log of USS *Tench* (SS 417), 1965:

The new year's log is a Navy tradition
And a poem it must be is the gruesome condition
If only a Shakespeare I were – there's the rub
Then this rhyme with iambics I'd gracefully dub.

Lt. G. W. Werenskjold

Other midwatch poets embrace their role as bard and compose elegant verse that mimics famous poets of history. From Poe's The Raven and Longfellow's Paul Revere's Ride to the popular 'Twas the Night Before Christmas, some OODs have relished the opportunity to pit their skills for rhyme and meter against the masters' widely known works. More on this later.

Sometimes the reasoning for writing a log in verse is as simple as relieving the boredom of a four or eight hour watch on New Year's Day. As Lt. Cmdr. L. F. Brooks, USS *George C. Marshall* (SSBN 654), 1969 writes

It's the custom of the service,
Though it makes some people nervous,
To write the New Year midwatch log in rhyme.
When the watch is long and boring,
And you'd rather be home snoring,
Making verses sometimes helps to pass the time.

Bringing a log entry to an end usually involves the invocation of a "Happy New Year" but often is accompanied by one last shot at softening the reader's expectation of a competent poem.

> Now that I've written
> All that facts I must scribble.
> With a "Happy New Year"
> I will end all this dribble.
>
> B. L. Doggett, Jr., LT j.g., USN
> USS *Sea Leopard* (SS 483), 1956

*Complaints about lack of alcohol and the companionship of women.* On July 1, 1914, ships in the US Navy became a desert with respect to the availability of alcoholic beverages. General Order No. 99 stated, "The use or introduction for drinking purposes of alcoholic liquors on board any naval vessel, or within any navy yard or station, is strictly prohibited, and commanding officers will be held directly responsible for the enforcement of this order." Consequently, when a holiday noted for its prodigious consumption of spirits rolled around, those on-board ship were left high and dry. There is no shortage of complaints about this state of affairs contained in New Year's Day logs. Frequently, in the same poem another common complaint arose. Midwatch poem writers bemoaned the fact that they were separated from their wife or girlfriend or were unable to fraternize with women in port. There was no New Year's kiss for those on watch.

Some writers used one of the older phrases related to alcohol consumption that dates to the days of ships with sails. If one of the most important lines in the rigging of a ship (the mainbrace) was cut, crew members had to repair the line by splicing. This sometimes had to be done while under fire which made the challenging job even more difficult. The crew who performed the duty were given a ration of alcohol as a reward. Splicing the mainbrace became a euphemism for consuming alcohol. One of the more famous users of the term was Medal of Honor recipient Eugene Fluckey who began a practice of doling out alcohol to the crew of USS *Barb* (SS 220) when she was able to sink a Japanese ship. Here are some representative example responses to the prohibition-like conditions on board including a warning about consuming too much of a good thing.

USS *Snook* (SSN 592), 1962

The duty section its vital watches stands
With heavy heart and disappointed face
Regretting the lost chance to join all hands
In a New Year's soiree' to splice the-main-brace.

P. J. Vogelberge, Jr., Lt, USN

USS *Caiman* (SS 323), 1959
Oh for a "scotch" or a can of cold beer
Instead of black coffee to bring in the New Year;
But we'll patiently wait 'till we get to Yokohama
For an overdue "hot bath" and much "Akadama"
[Note: Akadama is a sweet fortified Japanese wine]

R. M. Smith, Lt, USN
USS *Swordfish* (SSN 579), 1959

So Happy New Year all you lugs,
You alcoholic jitter bugs;
Just keep in mind that we feel swell
While your heads ache to beat all hell.

D. G. Smith, Lt, USN

Poem writers often bemoaned both the dearth of alcohol and the absence of their wives, girlfriends, or women in general. Lt. H. Sweet on *Argonaut* attempts to persuade himself that these deprivations work in his favor while Lt. j.g. Leike, Jr., aboard *Bream*, tries to cope with only a dream.

USS *Argonaut* (SS 475), 1961

Our crew is returning
Their noses are red.
They smell like Chanel
As they fall into bed

But I have the duty
And my thoughts – they are clear
They are not clouded
By whiskey or beer.
No women to distract me.
Nothing leads me toward sin
What a helluva fix
On New Years Eve – to be in.

H. Sweet, Lt., USN
USS *Bream* (SSK 242),1965

While out in Vallejo I hear the crew's happy shouts,
As champagne and beer pour forth from their spouts.
Alas, the various services received from the pier
Contain no beautiful women, not one ounce of cheer.
So it's early to bed and keep a taut watch
As I wish all "Happy New Year," and dream about Scotch.

H. W. Leike, Jr., Lt. j.g., USN

Lt. J. S. Lyman, Jr. brags about the amorous prowess of *Ronquil's* crew but suggests that SOPA may have something to say about fraternizing with the local women,

USS *Ronquil* (SS 396), 1959

So mothers beware – as lovers we're famous.
Keep your girls from the city, for their wits are no match
For the sailors who wear the Ronquil arm patch
On the island of North at a Place called NAS
There resides an admiral who can save your lass.

[Authors' Note: Lt. Cmdr. John S. Lyman, Jr. remains on eternal patrol as a crewmember of USS *Thresher* (SSN 593), lost on April 10, 1963]

But Lt. R. E. Hereford, Jr., suggests that consorting with local women is nevertheless occurring:

USS *Flasher* (SSN 613), 1969

The ship receiving services from ashore
While the off-duty sections enjoy wahine amour.

Lt. E. C. Baldwin urges his Sea Leopard mates to have a great time, but issues a warning:

USS *Sea Leopard*(SS 483), 1962

I figure that since I gotta be here
I'll think of my buddys over having a beer
So kiss the girls and drink up hearty
And toast the duty section who missed the party
And as I write this log in rhyme
Remember
It may be you with this duty next time.

*Mimicking Popular Poems.* As noted earlier, midwatch poets often record their log entries in the style of well-known or popular poems. In surface vessel logs, the most used styles include Poe's The Raven and the ever-popular 'Twas the Night Before Christmas. A few ventured into less-traveled territory, yielding poems in the popular "beat" style of the 50's and 60s. One poet expertly wove his log into Hamlet's soliloquy, beginning with "To sea, or not to sea, that is the question" while another fulfilled his log-keeping duty with a series of haiku. Submariners also seemed to possess an affinity for 'Twas the Night Before Christmas. A standout among the submarine midwatch poems, though, appears in the 1960 log from USS *Trout* (SS 566) by Lt. j.g. J. S. Hellewell, who mimics T. S. Elliot's The Love Song of J. Alfred Prufrock.

LET US GO THEN, YOU AND ME
WHEN TROUT IS LAID OUT IN CHARELSTON'S SEA
LIKE A FISH EMBALMED IN FORMALDEHYDE
MOORED AT THE STARBOARD SIDE
TO SS FIVE SEVEN SIX USS DARTER
IN A NEST OF TWO. AS TWO MARTYRS
WITH APPROPRIATE AND MEET
UNITS OF THE ATLANTIC FLEET
IS COMMANDER MINE LANT, SIR

THE LIEUTENANT'S WATCH IS LONG
HE LISTENS TO THE SIREN'S SONG
AS HE HEARKENS DOWN THE BEACH
THE MERMAIDS SINGING, EACH TO EACH
HE DOES NOT THINK IT IS TO HIM THEY SING
AS HE WATCHES COLD, UNSEEN

THE STARS FLICKER BRIGHT MIRAGES
OF A THOUSAND DISTANT IMAGES.
THE YOUNG LIEUTENANT HALF-RECALLS
GAY MUSIC AND GLOW OF NEW YEAR'S BALLS.

EXCITED LAUGHTER, HOLIDAY SOUND,
TOASTS RAISED HIGH, ROUND AFTER ROUND

BUT CHRISTMAS LEAVE IS BUT A MEMORY
THE YOUNG LIEUTENANT DRIFTS INTO REVERIE
IT IS MIDNIGHT; WITHOUT A SIGN
THE OLD YEAR PASSES INTO TIME.

IN THE COLD DECEMBER NIGHT
THE SHIP IS DESERTED, QUITE
SOLEMN AND ALONE, NO SOUND
OF MERRYMAKING WILL RESOUND
THROUGH THE DARK HOURS OF CHEER
THE COMING OF THE INFANT YEAR

THE MERMAIDS SING UPON THE BEACH
HE HEARS THEM SINGING, EACH TO EACH
BUT AS HE WATCHES, COLD UNSEEN,
IT IS NOT TO HIM THEY SING.

## Midwatch poems as reflections of the social and political milieu

Many midwatch poems reflect world political and social conditions of the post-WWII era. Cold War tensions, nuclear proliferation, the space race, and proxy wars around the globe take center stage in many New Year's logs. In 1962, Lt. j.g. James R. Burnett, aboard USS *Spikefish* (SS 404), pens one of the most eloquent discourses about the West's fears of the Soviet threat.

[Preceded by ship information]

For we are vigilant here tonight
As vigilant we must be.
To guard against that menace red
That has come to our southern sea.

This menace red that creeps and craws
And slithers in our path,
Has eaten up our neighbor south
And that slaketh not it's rath

For this catastrophic iconoclast
Would masticate the earth
Then turn and feed itself again
On those who gave it birth.

Appease it not nor approbate
It knows no honored word.
But uses trust and promise false
To mesmerize the herd.

But shotted gun it understands
And steel at highest port
The malled fist makes it stand at bay
And pay differential court.

So set the watch my shipmates bold
And keep it taunt and well.
And squarely face this manace red
This excrement from hell.

Do your duty, do it well
With resolution grand
And we will beat this menace red
And keep it from our land.

Lt. K. S. Porter, USS *Medregal*(SS 480), 1959, sums up the changing political and social conditions around the world and offers some reasons for these changes, along with some words of hope. After several stanzas of typical ship's conditions reports, Porter writes

> The call for this log, in verse writ so fine,
> Is the very first hour of the year fifty-nine.
>
> The end of the year of our Lord fifty-eight,
> Generates thought of some deeds that were great.
>
> The end of the world did not come as forecast,
> Nor even the submarine without any mast.
>
> The gain of our sciences seem so enormous,
> Our space deeds assure us the Russians won't swarm us.
>
> The rise of football, the hoop, and Van Cliburn
> More TV, more sports cars, they both make the eye burn.
>
> Yet history may find that the largest effect,
> Judged many years hence when the view is silent.
>
> Is not that of merchandise, glittering and neat,
> Nor even our missiles, nor nuclear fleet.
>
> But rather the growth in the minds of us all
> Of the seeds of a movement of stature so tall.
>
> If war is defined as a contest of wills,
> Effectiveness hinges on needs it fulfills.
>
> Morality, spirit, intensity, honor,
> Though technically strong, without we're a goner.
>
> Though it seems that a stalemate of arms is before us,
> This kindling of chances for greatness can soar us,
>
> Thus pointing for all of the nation to see,
> How men can improve when on needs they agree.
>
> The impact inherent makes the big picture clear:
> That the ideological war we not fear,

Since this shining example for the best of all worlds
Abundantly shows as this movement unfurls.

So here's to the start of another new year,
Just bristling with reasons for all to take cheer!

In a time of great apprehension, Lt. Porter was able to find room for optimism and hope.

### Do Submarine Midwatch Poems Differ from Surface Vessel Poems?

New Year's Day midwatch poems appearing in submarine deck logs show a few differences from those of surface vessels. Two clear examples from the opening lines of midwatch poems contrast submarines' most obvious differences

From USS *Cubera* (SS 347) in 1959, Lt. j.g. J. W. Smith writes

The Seawolf's record it might be fine,
But we submerged in '58 and came up in '59.
That may be a record some may say,
To us in Task Group Alfa, it's just another day.

While in that same year, Lt. j.g. L. L. Ingram aboard USS *Sterlet* (SS 392) begins with

From the ocean depth we arose,
To see what the hell goes.

These expressions clearly differentiate submarines from most other vessels. No surface ship sailor wants to contemplate the experience of being submerged! As poet Siew David Hii puts it, "Every ship gets to be a submarine **once**."[4] [Authors' emphasis]

Given that these logs were entered during some of the hottest periods of The Cold War, and given the secrecy involved in their operations, it is not surprising that submarine logs often reflected action during special operations. Brevity of log entries often reflected this situation with a simple "underway on special operations," but at least one OOD still used the occasion to spin a rhyme with

> SUBMERGED ON SPECIAL OPERATIONS,
> KEEPING PEACE AMONG THE NATIONS.
>
> G. S. Murphy, Lt. USN
> USS *Abraham Lincoln* (SSBN 602)
> January 1, 1962

This succinct verse may be the shortest rhyming midwatch poem found from any type of vessel.

Another notable difference between surface vessel midwatch poems and poems by the wearers of dolphins involves mentioning crewmembers by name. Although a few poets on surface ships mention the names of the ship's captain or XO, with the occasional addition of a quartermaster or bos'n, several submariners went to great lengths to mention more shipmates. The 1965 log of the USS *Capitaine* (AGSS 336) mentioned 84 specific crew names, which is probably the entire sub's complement. LT J. L. Potter began the recitation with the captain's name and the name of the most junior crew member, then went on to incorporate the wardroom, chiefs, and other crewmembers, mostly by section. Other submarine log entries that mention multiple crew members by name include:

USS *Bonita* (SSK 3) in 1958 mentions 47 crewmembers; USS *John Adams* (SSBN 620) 1968 mentions 38; and USS *Seawolf* (SSN 575) 1958 mentions 27. [See Figures 1 and 2]

It is difficult to know whether this penchant for listing crew members by name is due to the smaller crew size as compared to many surface vessels, enhanced crew cohesion resulting from the unique physical constraints of subs, or some other factor.

One final difference seen in submarine midwatch poems involves their physical structure. Sailors from all types of vessels used a variety of methods to present their poetic efforts. Some merely write or type their poem upside down on the pages of the log. Two use red ink. One poem cleverly pictures a ship using what can only be described as word art. Yet others express their creativity by reaching back over hundreds of years to record their midwatch poem using calligraphy and/or illumination. Lt. j.g. Charles E. Johnson, aboard USS *Atule* (SS 403), penned the midwatch in an Old English-type style in both 1961 and 1962 [See Figure 3]. Lt. John W. Roberts, USS *Barbel* (SS 580) in 1961 was the only poet who used a classic Medieval first initial or drop cap to begin his log [See Figure 4]. Of thousands of logs reviewed, only eight use some type of calligraphic representation. Five of those are submitted from subs.

Nevertheless, New Year's Day midwatch poems aboard US Navy submarines resemble those entered into the logs of surface vessels in most ways, with only a few interesting differences. Regardless of vessel type, poem writers likely assume their log in verse will be relegated to the anonymity of millions of pages of deck logs archived as official ship records. What would the writers of these poems think about them being read by a much larger audience than the ship's captain, navigator, quartermaster, and a few others? A few poems in the 1960s reached a wider audience when *All Hands*, an official US Navy publication, printed a few judged to be the best of the previous year. But for the most part, few people saw them, and the writers were probably just happy to be done with their bardic task. Many authors' sentiments likely mirrored those of Lt. j.g. J. M. Leeds aboard USS *Sirago* (SS 485) in 1962.

> We're steaming in circles with batteries chargin'
> (If you can rhyme better make notes in the margin)
> Here's to the lady in the Pentagon cellars
> Who reads the logs of us sea-going fellers
> If you don't like our poem don't make a fuss
> Just don't send the deck log back to us.

## Figures

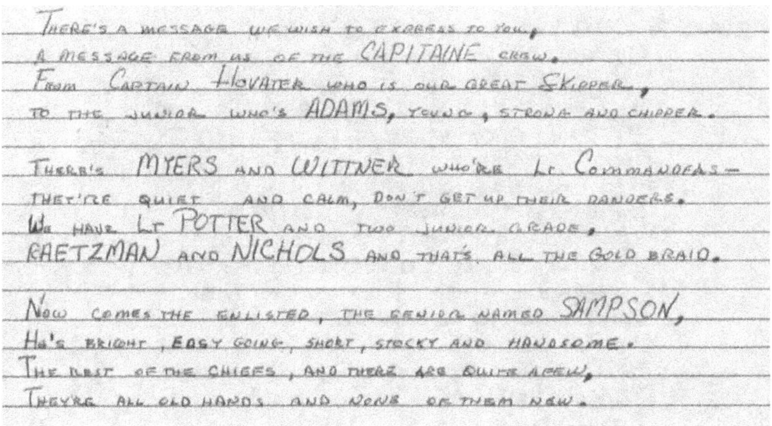

THERE'S A MESSAGE WE WISH TO EXPRESS TO YOU,
A MESSAGE FROM US OF THE CAPITAINE CREW.
FROM CAPTAIN HOVATER WHO IS OUR GREAT SKIPPER,
TO THE JUNIOR WHO'S ADAMS, YOUNG, STRONG AND CHIPPER.

THERE'S MYERS AND WITTNER WHO'RE LT COMMANDERS—
THEY'RE QUIET AND CALM, DON'T GET UP THEIR DANDERS.
WE HAVE LT POTTER AND TWO JUNIOR GRADE,
RAETZMAN AND NICHOLS AND THAT'S ALL THE GOLD BRAID.

NOW COMES THE ENLISTED, THE SENIOR NAMED SAMPSON,
HE'S BRIGHT, EASY GOING, SHORT, STOCKY AND HANDSOME.
THE REST OF THE CHIEFS, AND THERE ARE QUITE A FEW,
THEY'RE ALL OLD HANDS AND NONE OF THEM NEW.

Figure 1. From USS *Capitaine* (AGSS 336),
National Archives, NAID: 212626474,
Container ID: 683, January 1, 1965

NEXT FOR YOUR PLEASURE, THE TM'S YOU SHOULD GREET—
HALDEMAN, THOMPSON AND MILLER YOU SHOULD MEET.
PEÑA, SASSENGER AND MACKELPRANG ARE ALL HERE
AND GRANDFATHER BRUNDAGE IN HIS 60TH YEAR.

AND NOW FOR OUR COOKS, WHO'RE REAL NECESSARY,
WE HAVE COLAVECHIO, CARR AND ALSO NERI.
FOR STEWARDS WE HAVE MERCADO AND DUPLON,
GUTIERREZ AND CADACIO OF WHOM WE'RE REAL PROUD.

Figure 2. From USS *Capitaine* (AGSS 336), National Archives, NAID: 212626474, Container ID: 683, January 1, 1965

00-08 In ARD Twelve we rest tonight
On keel blocks firm and shored up tight
This New Years night for us goes hard
At Charleston, S.C. ; U.S. Naval Shipyard
Those vessels present are sharp and neat
Atlantic units of our U.S. Fleet
Foreign vessels we'll see this day
You'll find them listed by SOPA
Yard and District craft are here
Moored hard and fast each to their pier
In chain of command there's no new slant
Our SOPA is COMINLANT
Services received - none extraneous
ARD Twelve provides - all miscellaneous
The whistles blow and sirens blast
And Sixty-One is here at last.

APPROVED: EXAMINED: C. E. Johnson
LTJG, USNR

Figure 3. From USS *Atule* (SS 403), National Archives, NAID: 121602883, Container ID: 247, January 1, 1961 (Authors' note: "ARD Twelve" was a floating drydock)

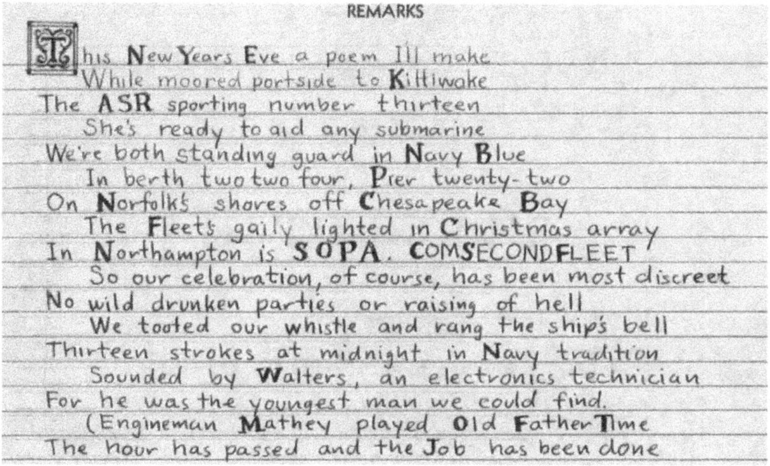

REMARKS

This New Years Eve a poem Ill make
While moored portside to Kittiwake
The ASR sporting number thirteen
She's ready to aid any submarine
We're both standing guard in Navy Blue
In berth two two four, Pier twenty-two
On Norfolk's shores off Chesapeake Bay
The Fleets gaily lighted in Christmas array
In Northampton is SOPA. COMSECONDFLEET
So our celebration, of course, has been most discreet
No wild drunken parties or raising of hell
We tooted our whistle and rang the ship's bell
Thirteen strokes at midnight in Navy tradition
Sounded by Walters, an electronics technician
For he was the youngest man we could find.
(Engineman Mathey played Old Father Time
The hour has passed and the Job has been done

Figure 4. From USS *Barbel* (SS 580), National Archives, NAID: 182714280, Container ID: 289, January 1, 1961.

# SIDELIGHT 11

## The Long and Short of It!

What are the shortest and longest midwatch poems? Given the high frequency of complaints about writing the midwatch in verse, it should not be surprising that some OODs chose brevity for their log entry. Others chose to write short poems because they were on special operations and were limited in what they could enter in the log.

Three poems of nearly equal length competed for the shortest poem designation. That honor went to the poem cited earlier by Lt. G. S. Murphy aboard USS *Abraham Lincoln* (SSBN 602) in 1962.

> Submerged on special operations,
> Keeping peace among the nations.

Two other poems are close in brevity to Murphy's. Lt. j.g. H. K. Graf, aboard USS *Diphda* (AKA 59) while moored at Todd Shipyard in Alameda in 1956, bemoaned pulling midwatch duty with

> Todds – Alameda,
> Moored dock 2 E.

New Years Eve
Got duty – phooey.

Ensign L. A. Hartley also entered a very short poem into the log of USS *LSM* 297 in 1959.

Ring out the old, bring in the new,
Here's to a Happy New Year to you.

However, the ensign also added several lines of traditional log entries making the midwatch log much longer.

Lt. j.g. C. H. Malone aboard USS *Currituck* (AV 7) in 1962 chose to write a rhyme that included events in the early history of the ship as well as the year just ended. The poem spanned nearly five typewritten pages and is the longest in our database.[1] The poem is extraordinary in that the author maintained a consistent meter throughout its considerable length. It begins with:

Please listen good friends and a tale I'll recite
(Seems a good thing to do on the mid-watch tonight)
About a good ship and the seas that she sailed;
I'll tell you all this all before morning has paled.

She's called the Wild Goose by her Captain and crew;
She's also called others, but it just wouldn't do
To tell you those names; that's not my intention.
'Tis best in a poem those names not to mention.

Currituck is her name, Philadelphia made,
World War II her vintage, seaplane tending her trade.
The Atlantic, Pacific she's sailed without fear,
Taking care of her brood and all of their gear.

Twice placed in commission, twice called on to serve,
And service like that takes plenty of nerve.
The third time a charm, so the old saying goes;
This saying is true, as most everyone knows.

.................and ends with

Section Two has the duty, alert and on time.
To watch o'er the Goose and compose this quaint rhyme;
But the time draws nigh when reliefs do appear,
The first watch has ended as we start this year.

You've read this, I hope, from beginning to end;
If so, we are sure we can call you our friend.
But let us not linger, 'tis time to secure
Sixty-two verses is enough to endure.

But you've heard of the Goose, you know of her fame,
And few that have known her are ever the same.
To greet you this season of gladness and cheer —
From the Wild Goose's crew — A Happy New Year!

.............and that is the long and short of it!

# CHAPTER 11

## Graphic Embellishments: How Design Added Beauty to Deck Log Poems

Writing a midwatch deck log poem on New Year's Day allows Officers of the Deck (OODs) to express their creativity by writing the log in verse. As seen in chapter 1, poems contain complaints about being compelled by tradition to complete this task. It seems that merely writing a poem would be sufficient to quell the OODs' consternation. Yet, some poem writers chose to embellish their bardic efforts beyond simply writing verse in standard lines and stanzas. These additional embellishments were sometimes completed by simple alterations of log protocol. Others required considerable effort on the part of the writers.

Of over 10,000 midwatch poems reviewed by the Midwatch In Verse Project members, two were written in red ink. USS *Yorktown* (CVA 10, 1965) and USS *Embattle* (MSO 434, 1967) logs were brightened by crimson text, possibly a carryover of a color associated with Christmas just a week prior. Both log entries were signed by their writers in red ink, which may have violated official protocol that dictated black or blue ink.

Four logs contained poems that were written upside down with respect to the log's page orientation. Interestingly, three of the four were from the same ship, USS *Stormes* (DD 780) in 1962, 1963, and 1966. In a similar vein, Ensign F. W. Trumpler aboard USS *Rushmore* (LSD 14, 1958) signed his name backwards, KCIREDERF MAILLIW RELPMURT.

One industrious officer aboard USS *Etlah* (AN 79), a Cohoes-Class net laying ship, managed to write a short poem that contained no important ship information. His poem simply described his log entry that consisted of a detailed map complete with legend. Lt. j.g. C. T. Berger deftly drew a map of Subic Bay, Philippines, showed *Etlah's* position on the map, and added a wish of "Happy New Year" within a series of bells (See Figure 1). Berger's log entry is truly one of a kind.

Also one of a kind is Lt. j.g. B. W. Young's poem that depicts a ship's silhouette reminiscent of the ASCII art of the late 1970s and the 1980s when dot matrix printers became available. Presumably, the log depicts USS *Parsons* (DDG 33), a Forrest-Sherman class destroyer that was converted from an all-gun vessel to a guided missile destroyer (See Figure 2).

One of the more common embellishments found in midwatch poems was the acrostic. In poetry, an acrostic occurs when the writer uses the first letter in each line to spell out a word or phrase when read vertically. More than a dozen examples appeared in the 10,000+ poems reviewed from the 1950s through 1970s. Most acrostic poems appeared in logs during the 1960s.

Two types of acrostics dominated this category of embellished poems. Wishing a "Happy New Year" or some representation of the ship's name (sometimes including its designation and hull number) appeared most frequently. A few writers managed a combination of both (See Figures 3 and 4). Two acrostic poems deserve special mention. In 1967, while aboard USS *Wandank* (ATA 204), Warrant Officer K. A. Flynn used the midwatch poem to spell out *Håfa adai*, a Chamorro greeting like "Aloha" in Hawaii. *Wandank* spent much of her time operating out of Guam and Flynn took the opportunity to offer the reader a greeting familiar to the regional culture.

A more curious acrostic appeared in the 1965 log of USS *Hank* (DD 702). The acrostic "PROJECT SCRAP" had no apparent referent within the poem. So, what was this cryptic message being sent by the writer, Lt. j.g. Bruce G. Whalen? *Project SCRAP* was a US Navy initiative to cut down on wasteful paperwork. SCRAP stands for *Selective Curtailment of Reports and Paperwork* and was supported by the upper echelons of the Department of the Navy. The project aimed to enhance readiness by reducing paperwork. As some critics pointed out, however, the very paperwork that the Navy floundered under presumably existed to enhance readiness in the first place. The irony was not lost on many Navy personnel. Lt. j.g. Whalen may have used his midwatch poem as a personal protest of generating a log entry that had little to do with his ship's battle readiness (See Figure 5).

Some poets reached back centuries to add a creative touch to their poetry by employing calligraphy and/or illumination. Calligraphy is the art of beautiful writing, and it has been practiced for millennia in the Near and Far East. In the West, the terms *calligraphy* and *illumination* tend to evoke images of beautiful Medieval manuscripts written in fonts with artistic flourish and often accompanied by pictorial representations of the sacred and mundane of the times. The illumination of documents involves the addition of pictures or borders that serve to enhance the beauty of calligraphy.

Often the drawings appeared in the margins of the text and were referred to as marginalia.

Some midwatch deck log poets demonstrated a degree of skill in the art of calligraphy. An excellent example of this ability appears in the log of submarine USS *Atule* (SS 403). Lt. j.g. Charles E. Johnson used a script similar to blackletter/Gothic that was popular in Western Europe in the 12th through 18th centuries (See Figure 6). Johnson also used the same script when he wrote the midwatch log aboard *Atule* the previous year, 1961. Notably, many poems employing calligraphic techniques appeared in the logs of submarines.

A small number of midwatch poets demonstrated their skills of manuscript illumination. In a striking example of the addition of graphics to a deck log, Ensign J. M. Chouinard created a modified acanthus motif with anchors to decorate the midwatch log of USS *Ogden* (LPD 5) in 1968 (See Figure 7). In a possible case of wish fulfillment, Lt. j.g. C. P. Phillips, USS *John Hood*, (DD 655) added delightful marginalia depicting the results of what appears to be the aftermath of joyous New Year's celebration (See Figures 8 and 9). No such celebration would have been possible for those who had the watch.

Finally, in the January 1, 1956, log of USS *Willard Keith* (DD 775) an unknown sailor reproduced the ship's patch designed by cartoonist Hank Ketchum. The graphic (see Figure 10) precedes the list of ship's officers in the log and features Dennis the Menace attacking an enemy submarine with a hammer, with the caption "Menace of the Sea."

The midwatch is often considered to be boring and lonely duty. These examples provide evidence of the lengths to which sailors go to alleviate those feelings and bring some flair to their deck log entries.

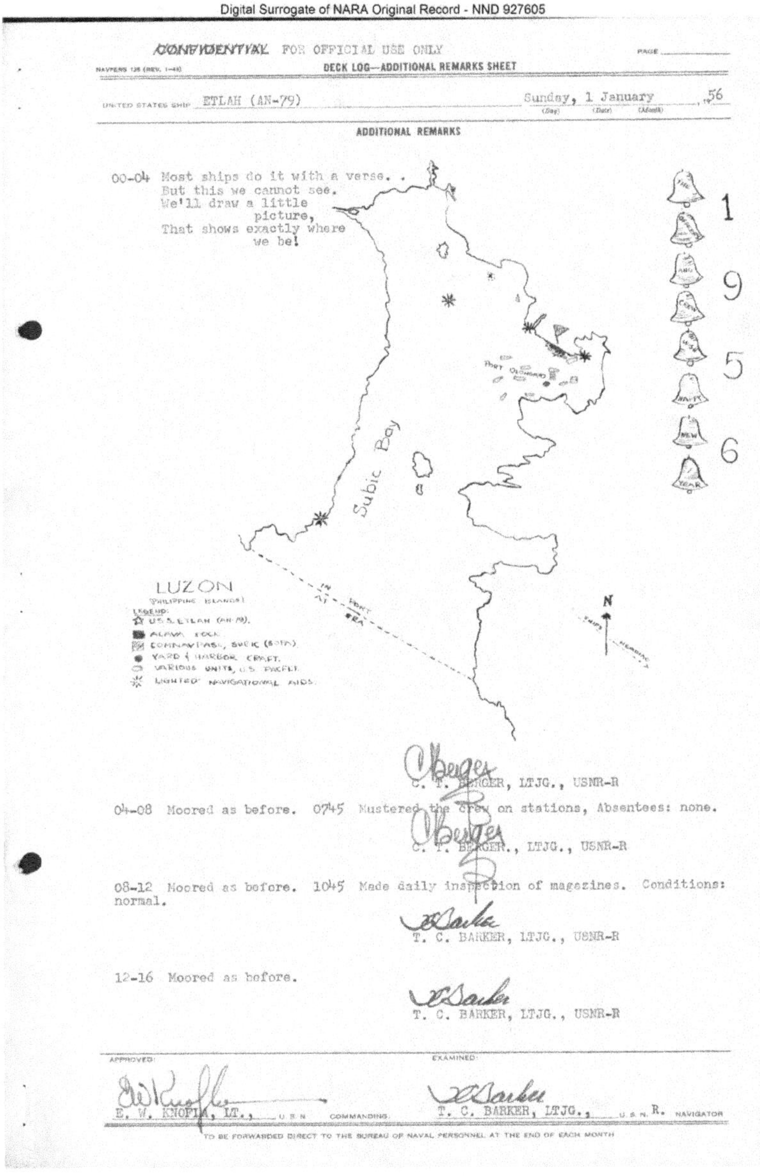



~~CONFIDENTIAL~~ FOR OFFICIAL USE ONLY PAGE ____

NAVPERS 135 (REV. 1-48) DECK LOG—ADDITIONAL REMARKS SHEET

UNITED STATES SHIP ETLAH (AN-79) Sunday, 1 January, 1956
(Day) (Date) (Month)

ADDITIONAL REMARKS

00-04 Most ships do it with a verse. .
But this we cannot see.
We'll draw a little
picture,
That shows exactly where
we be!



C. T. BERGER, LTJG., USNR-R

04-08 Moored as before. 0745 Mustered the crew on stations, Absentees: none.

C. T. BERGER., LTJG., USNR-R

08-12 Moored as before. 1045 Made daily inspection of magazines. Conditions: normal.

T. C. BARKER, LTJG., USNR-R

12-16 Moored as before.

T. C. BARKER, LTJG., USNR-R

APPROVED: E. W. KNOFLA, LT., U.S.N. COMMANDING

EXAMINED: T. C. BARKER, LTJG., U.S.N.R. NAVIGATOR

TO BE FORWARDED DIRECT TO THE BUREAU OF NAVAL PERSONNEL AT THE END OF EACH MONTH

Figure 1 From USS *Etlah* (AN 79), National Archives, NAID: 219879954, Container ID: 1427, January 1, 1956.

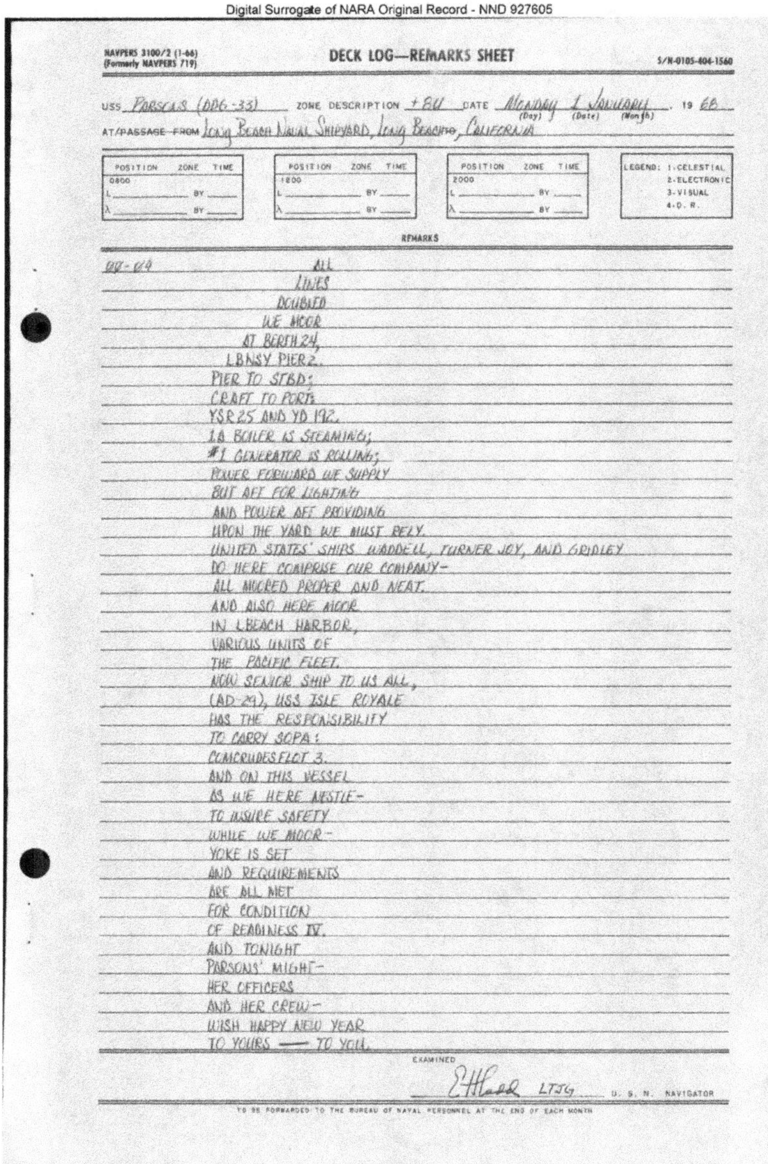
Digital Surrogate of NARA Original Record - NND 927605

NAVPERS 3100/2 (1-66)
(Formerly NAVPERS 719)

DECK LOG—REMARKS SHEET

S/N-0105-404-1560

USS PARSONS (DDG-33) ZONE DESCRIPTION +8U DATE MONDAY 1 JANUARY 19 68
(Day) (Date) (Month)

AT/PASSAGE FROM LONG BEACH NAVAL SHIPYARD, LONG BEACH, CALIFORNIA

| POSITION | ZONE | TIME |
|---|---|---|
| 0800 | | |
| L ____ BY ____ | | |
| λ ____ BY ____ | | |

| POSITION | ZONE | TIME |
|---|---|---|
| 1200 | | |
| L ____ BY ____ | | |
| λ ____ BY ____ | | |

| POSITION | ZONE | TIME |
|---|---|---|
| 2000 | | |
| L ____ BY ____ | | |
| λ ____ BY ____ | | |

LEGEND: 1-CELESTIAL
2-ELECTRONIC
3-VISUAL
4-D. R.

REMARKS

00-04 ALL
LINES
DOUBLED
WE MOOR
AT BERTH 24,
LBNSY PIER 2.
PIER TO STBD;
CRAFT TO PORT:
YSR 25 AND YD 192.
1A BOILER IS STEAMING;
#1 GENERATOR IS ROLLING;
POWER FORWARD WE SUPPLY
BUT AFT FOR LIGHTING
AND POWER AFT PROVIDING
UPON THE YARD WE MUST RELY.
UNITED STATES' SHIPS WADDELL, TURNER JOY, AND GRIDLEY
DO HERE COMPRISE OUR COMPANY-
ALL MOORED PROPER AND NEAT.
AND ALSO HERE MOOR
IN LBEACH HARBOR,
VARIOUS UNITS OF
THE PACIFIC FLEET.
NOW SENIOR SHIP TO US ALL,
(AD-29), USS ISLE ROYALE
HAS THE RESPONSIBILITY
TO CARRY SOPA:
COMCRUDESFLOT 3.
AND ON THIS VESSEL
AS WE HERE NESTLE-
TO INSURE SAFETY
WHILE WE MOOR-
YOKE IS SET
AND REQUIREMENTS
ARE ALL MET
FOR CONDITION
OF READINESS IV.
AND TONIGHT
PARSONS' MIGHT-
HER OFFICERS
AND HER CREW-
WISH HAPPY NEW YEAR
TO YOURS — TO YOU.

EXAMINED

E Hall LTJG U. S. N. NAVIGATOR

TO BE FORWARDED TO THE BUREAU OF NAVAL PERSONNEL AT THE END OF EACH MONTH

*Figure 2* From USS *Parsons* (DDG 33), National Archives, NAID: 173213177, Container ID: 3202, January 1, 1968.

NAVPERS 3100/2 (1-66)
(Formerly NAVPERS 719)

**DECK LOG—REMARKS SHEET**

S/N 0105-904-1020

USS BIGELOW (DD-942) ZONE DESCRIPTION +5R DATE THURSDAY (Day) 1 (Date) JANUARY (Month), 19 7Ø

AT/~~PASSAGE FROM~~ U.S NAVAL STATION, MAYPORT FLORIDA ~~TO~~

| POSITION | ZONE | TIME |
|---|---|---|
| 0800 | | |
| L ________ BY ____ | | |
| λ ________ BY ____ | | |

| POSITION | ZONE | TIME |
|---|---|---|
| 1200 | | |
| L ________ BY ____ | | |
| λ ________ BY ____ | | |

| POSITION | ZONE | TIME |
|---|---|---|
| 2000 | | |
| L ________ BY ____ | | |
| λ ________ BY ____ | | |

LEGEND: 1-CELESTIAL
2-ELECTRONIC
3-VISUAL
4-D. R.

REMARKS

ØØ-Ø4 HERE SHE LIES BERTHED IN DELTA THREE-TWO, AS PRETTY A SHIP AS EVER YOU'LL SEE.
AULT AND THE BAILEY TO STARBOARD ARE MOORED WITH BRUMBY TIED UP TWIXT US AND THE PIER.
PROPERLY MOORED WITH SIX LINES ALL NOW DOUBLED; FORE AND AFT WIRE'S OUT SECURE AS CAN BE.
PROBABLY HALF OF THE CREW AT THIS TIME ARE OUT ON THE TOWN ENJOYING THEIR BEER,
YET WE WHO REMAIN ARE BOOSTED BY THOUGHTS THAT WHEN THEY RETURN IT'S OUR TURN TO CHEER.
NAV STA AT MAYPORT'S OUR PRESENT LOCATION WITH VARIOUS OTHER FINE SHIPS OF THE FLEET.
EVEN DISTRICT AND YARD CRAFT ARE TIED UP HERE TOO; ALL CELEBRATING A HAPPY NEW YEAR.
WHO THEN IS SOPA, THAT SALTY OLD MAN, SENIOR TO ALL SHIPS IN MAYPORT YOU'LL MEET?
YES, NONE, BUT COMMANDER, CRUDESFLOT TWELVE, EMBARKED IN YOSEMITE NOW AT THIS TIME.
EACH YOKE TYPE FITTING HAS NOW BEEN SECURED; CONDITION OF READINESS FIVE HAS BEEN SET.
ALL IS NOW READY TO START THE NEW YEAR, WHICH EVERYONE HOPES WILL BRING PEACE TO MANKIND.
REMEMBER THEN PLEASE, LEST YOU FORGET, HAVE THE HAPPIEST HAPPY NEW YEAR YOU'VE HAD YET.

E. W. Augusta
E.W. AUGUSTA LTJG, USNR

Ø4-Ø8 MOORED AS BEFORE

E. W. Augusta
E.W. AUGUSTA LTJG, USNR

Ø8-12 MOORED AS BEFORE. Ø81Ø MUSTERED THE CREW ON STATION. ABSENTEES: DOBSON, B.W., B746294 SA, USN; KELLER, R.E., 5877319 MM2, USN UA SINCE Ø73Ø THIS DATE.

E. W. Augusta
E.W. AUGUSTA LTJG, USNR

12-16 MOORED AS BEFORE.

E. W. Augusta
E. W. AUGUSTA LTJG, USNR

16-2Ø MOORED AS BEFORE.

E. W. Augusta
E. W. AUGUSTA LTJG, USNR

2Ø-24 MOORED AS BEFORE. 2209 MALANDRI, R.E. B45 63 19, SA, USN RETURNED ABOARD HAVING BEEN UA SINCE 0730, 28 DEC 69

Eric W. Augusta
E. W. AUGUSTA LTJG, USNR

EXAMINED

U. S. N. NAVIGATOR

TO BE FORWARDED TO THE BUREAU OF NAVAL PERSONNEL AT THE END OF EACH MONTH

*Figure 3* From USS *Bigelow* (DD 942), National Archives, NAID: 6927184, Container ID: 421, January 1, 1970

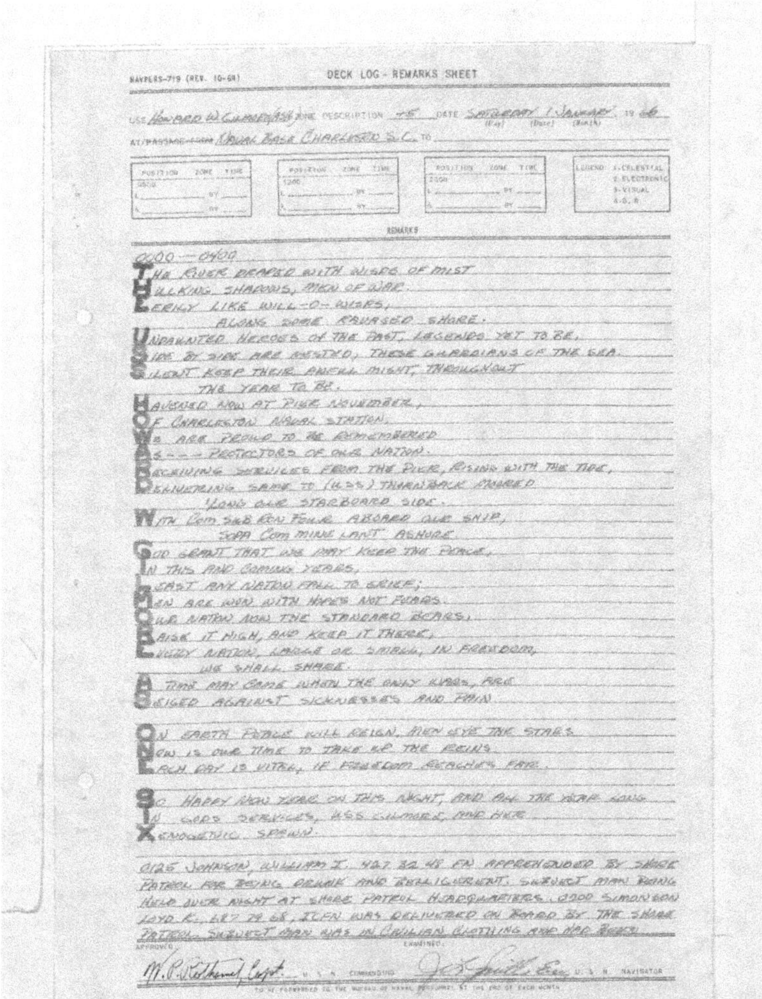

NAVPERS-719 (REV. 10-6?)

DECK LOG - REMARKS SHEET

USS Howard W. Gilmore (AS 16) ZONE DESCRIPTION +5 DATE Saturday 1 January 1966

AT/PASSAGE FROM Naval Base Charleston S.C. TO

REMARKS

0000 — 0400

THE RIVER DRAPED WITH WISPS OF MIST
HULKING SHADOWS, MEN OF WAR.
EERILY LIKE WILL-O-WISPS,
ALONG SOME [illegible] SHORE.
UNDAUNTED HEROES OF THE PAST, LEGENDS YET TO BE,
SIDE BY SIDE ARE NESTED, THESE GUARDIANS OF THE SEA.
SILENT KEEP THEIR [illegible] MIGHT, THROUGHOUT
THE YEAR TO BE.
HAWSERED NOW AT PIER NOVEMBER,
OF CHARLESTON NAVAL STATION.
WE ARE PROUD TO BE REMEMBERED
AS — PROTECTORS OF OUR NATION.
RECEIVING SERVICES FROM THE PIER, RISING WITH THE TIDE,
DELIVERING SAME TO (USS) THORNBACK MOORED
'LONG OUR STARBOARD SIDE.
WITH COM SUB RON FOUR ABOARD OUR SHIP,
SOPA COM MINE LANT ASHORE.
GOD GRANT THAT WE MAY KEEP THE PEACE,
IN THIS AND COMING YEARS,
LEST ANY NATION FALL TO GRIEF;
MEN ARE WON WITH HOPES NOT FEARS.
OUR NATION NOW THE STANDARD BEARS,
RAISE IT HIGH, AND KEEP IT THERE,
EVERY NATION, LARGE OR SMALL, IN FREEDOM,
WE SHALL SHARE.
A TIME MAY COME WHEN THE ONLY [illegible], ARE
[illegible] AGAINST SICKNESSES AND PAIN
ON EARTH PEACE WILL REIGN, MEN EYE THE STARS
NOW IS OUR TIME TO TAKE UP THE REINS
EACH DAY IS VITAL, IF FREEDOM REACHES FAR.
SO HAPPY NEW YEAR ON THIS NIGHT, AND ALL THE YEAR LONG
IN GODS SERVICES, USS GILMORE AND HER
ENERGETIC SPAWN.

0125 JOHNSON, WILLIAM I. 427 32 48 FN APPREHENDED BY SHORE PATROL FOR BEING DRUNK AND BELLIGERENT. SUBJECT MAN BEING HELD OVER NIGHT AT SHORE PATROL HEADQUARTERS. 0200 SIMONSON LLOYD K. 487 79 68, ICFN WAS DELIVERED ON BOARD BY THE SHORE PATROL. SUBJECT MAN WAS IN CIVILIAN CLOTHING AND HAD BEEN [illegible]

APPROVED: W. P. [illegible], Capt. U. S. N. COMMANDING

EXAMINED: [illegible], Ens. U. S. N. NAVIGATOR

TO BE FORWARDED TO THE BUREAU OF NAVAL PERSONNEL AT THE END OF EACH MONTH

*Figure 4* From *Howard W. Gilmore* (AS 16), National Archives, NAID: 12005505, [no Container ID listed], January 1, 1966

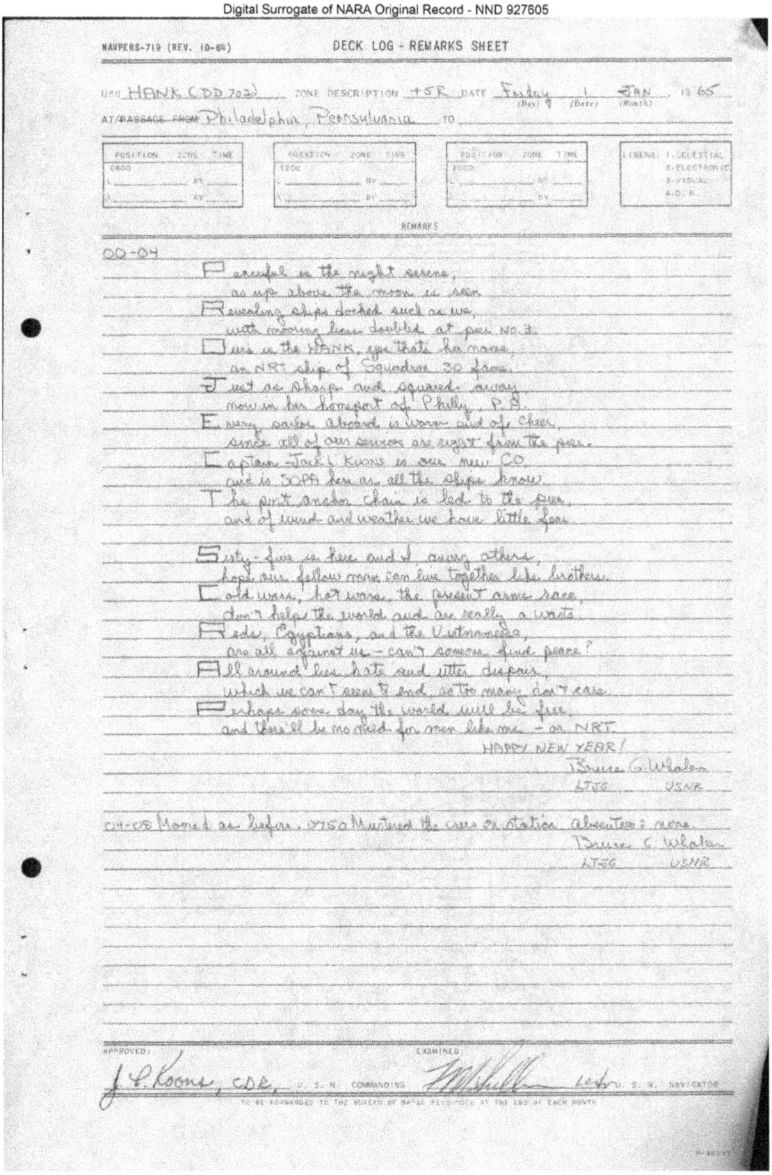

Digital Surrogate of NARA Original Record - NND 927605

NAVPERS-719 (REV. 10-64) DECK LOG - REMARKS SHEET

USS HANK (DD 702) ZONE DESCRIPTION +5R DATE Friday 1 JAN 1965

AT/PASSAGE FROM Philadelphia, Pennsylvania TO

REMARKS

00-04

Peaceful is the night serene,
as up above the moon is seen
Revealing ships docked such as we,
with mooring lines doubled at pier NO. 3.
Ours is the HANK, yes that's her name,
an NRT ship of Squadron 30 fame.
Just as sharp and squared away,
now in her homeport of Philly, P.A.
Every sailor aboard is warm and of cheer,
since all of our services are right from the pier.
Captain Jack L Koons is our new CO,
and is SOPA here as all the ships know.
The port anchor chain is led to the pier,
and of wind and weather we have little fear.

Sixty-five is here and I, among others,
hope our fellow man can live together like brothers.
Cold wars, hot wars, the present arms race,
don't help the world and are really a waste.
Reds, Egyptians, and the Vietnamese,
are all against us - can't someone find peace?
All around lies hate and utter despair,
which we can't seem to end, as too many don't care.
Perhaps some day the world will be free,
and there'll be no need for men like me - or NRT.

HAPPY NEW YEAR!
Bruce G. Whalen
LTJG USNR

04-08 Moored as before. 0750 Mustered the crew on station absentees: none.
Bruce G. Whalen
LTJG USNR

APPROVED: J. L. Koons, CDR, U. S. N. COMMANDING

EXAMINED: Lt. U. S. N. NAVIGATOR

TO BE FORWARDED TO THE BUREAU OF NAVAL PERSONNEL AT THE END OF EACH MONTH

*Figure 5* USS *Hank* (DD 702), National Archives, NAID: 214627405, Container ID: 1796, January 1, 1965.

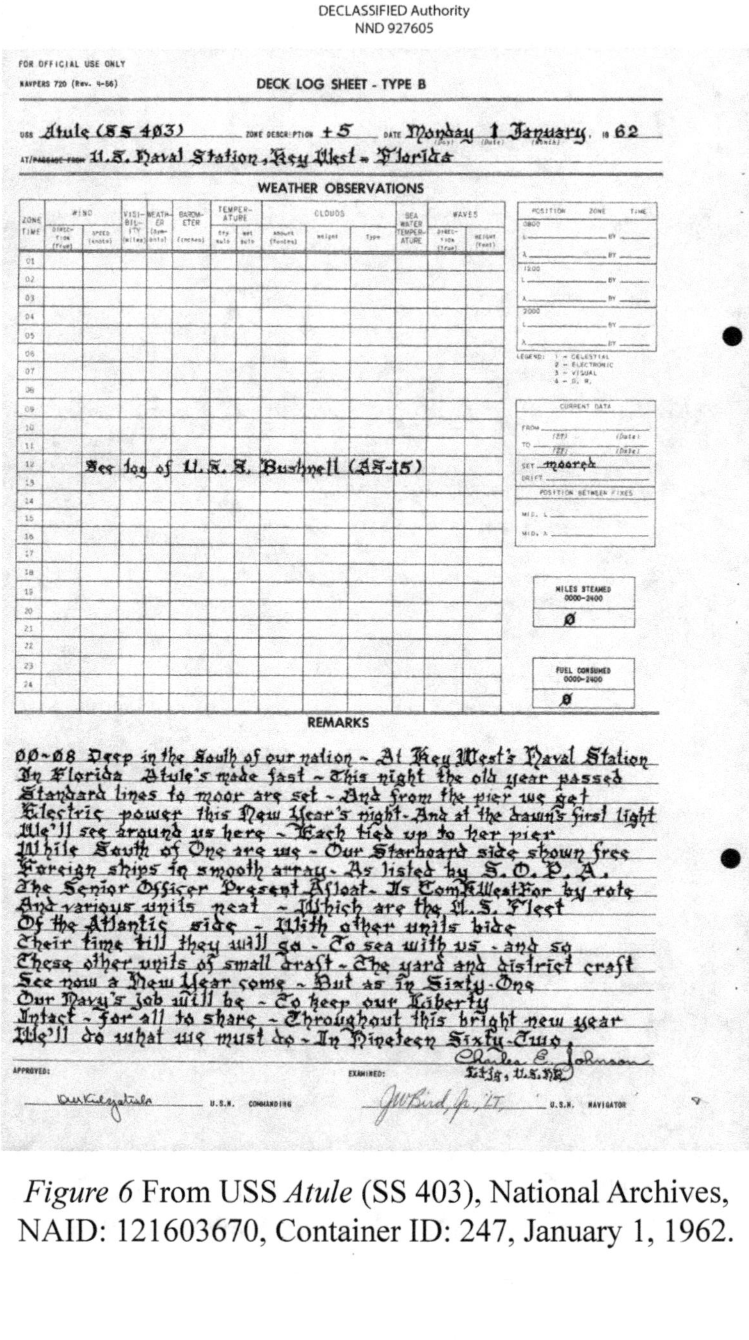

DECLASSIFIED Authority
NND 927605

FOR OFFICIAL USE ONLY
NAVPERS 720 (Rev. 4-56)

**DECK LOG SHEET - TYPE B**

USS Atule (SS 403) ZONE DESCRIPTION +5 DATE Monday 1 January, 19 62

AT/PASSAGE FROM U.S. Naval Station, Key West - Florida

**WEATHER OBSERVATIONS**

See log of U.S.S. Bushnell (AS-15)

CURRENT DATA — SET moored

MILES STEAMED 0000-2400: Ø

FUEL CONSUMED 0000-2400: Ø

**REMARKS**

ØØ-Ø8 Deep in the South of our nation - At Key West's Naval Station
In Florida Atule's made fast - This night the old year passed
Standard lines to moor are set - And from the pier we get
Electric power this New Year's night - And at the dawn's first light
We'll see around us here - Each tied up to her pier
While South of One are we - Our Starboard side shown free
Foreign ships in smooth array - As listed by S.O.P.A.
The Senior Officer Present Afloat - Is ComKWestFor by rote
And various units neat - Which are the U.S. Fleet
Of the Atlantic side - With other units bide
Their time till they will go - To sea with us - and so
These other units of small draft - The yard and district craft
See now a New Year come - But as in Sixty-One
Our Navy's job will be - To keep our Liberty
Intact - For all to share - Throughout this bright new year
We'll do what we must do - In Nineteen Sixty-Two.

Charles E. Johnson
Ltjg, U.S.NR

APPROVED: U.S.N. COMMANDING

EXAMINED: JW Bird, Jr., LT, U.S.N. NAVIGATOR

*Figure 6* From USS *Atule* (SS 403), National Archives, NAID: 121603670, Container ID: 247, January 1, 1962.

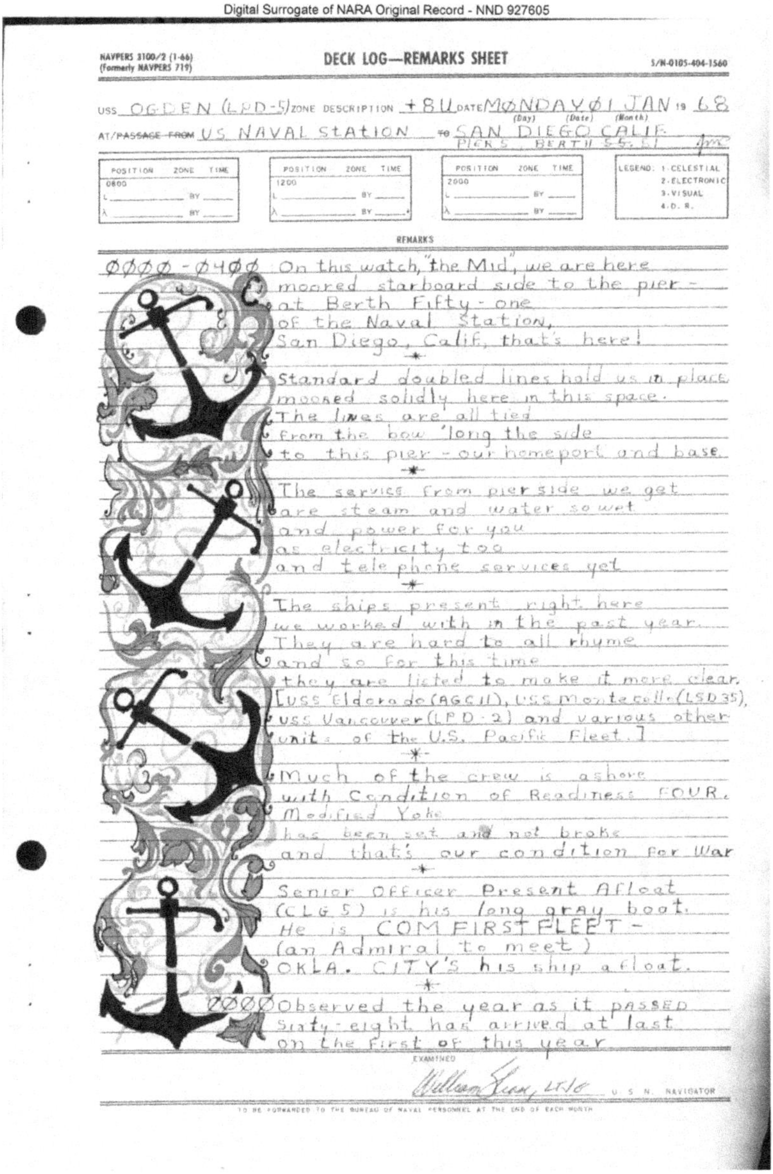

Digital Surrogate of NARA Original Record - NND 927605

NAVPERS 3100/2 (1-66)
(Formerly NAVPERS 719)

DECK LOG—REMARKS SHEET

S/N-0105-404-1560

USS OGDEN (LPD-5) ZONE DESCRIPTION +8U DATE MØNDAY Ø1 JAN 19 68
(Day) (Date) (Month)

AT/~~PASSAGE FROM~~ US NAVAL STATION ~~TO~~ SAN DIEGO CALIF.
PIERS, BERTH 55, 51

| POSITION | ZONE | TIME |
|---|---|---|
| 0800 | | |
| L ______ BY ______ | | |
| λ ______ BY ______ | | |

| POSITION | ZONE | TIME |
|---|---|---|
| 1200 | | |
| L ______ BY ______ | | |
| λ ______ BY ______ | | |

| POSITION | ZONE | TIME |
|---|---|---|
| 2000 | | |
| L ______ BY ______ | | |
| λ ______ BY ______ | | |

LEGEND: 1-CELESTIAL
2-ELECTRONIC
3-VISUAL
4-D. R.

REMARKS

ØØØØ - Ø4ØØ On this watch, "the Mid", we are here
moored starboard side to the pier -
at Berth Fifty-one
of the Naval Station,
San Diego, Calif. that's here!

*

Standard doubled lines hold us in place
moored solidly here in this space.
The lines are all tied
from the bow 'long the side
to this pier - our homeport and base.

*

The service from pierside we get
are steam and water so wet
and power for you
as electricity too
and telephone services yet

*

The ships present right here
we worked with in the past year.
They are hard to all rhyme
and so for this time
they are listed to make it more clear.
[USS Eldorado (AGC 11), USS Monticello (LSD 35),
USS Vancouver (LPD-2) and various other
units of the U.S. Pacific Fleet.]

*

Much of the crew is ashore
with Condition of Readiness FOUR.
Modified Yoke
has been set and not broke
and that's our condition for War

*

Senior Officer Present Afloat
(CLG 5) is his long gray boat.
He is COM FIRST FLEET -
(an Admiral to meet)
OKLA. CITY'S his ship afloat.

*

ØØØØ Observed the year as it PASSED
Sixty-eight has arrived at last
on the First of this year.

EXAMINED

William Leary, LT/JG U. S. N. NAVIGATOR

TO BE FORWARDED TO THE BUREAU OF NAVAL PERSONNEL AT THE END OF EACH MONTH

*Figure 7* From USS *Ogden* (LPD 5), National Archives, NAID: 173471260, Container ID: 3096, January 1, 1968.

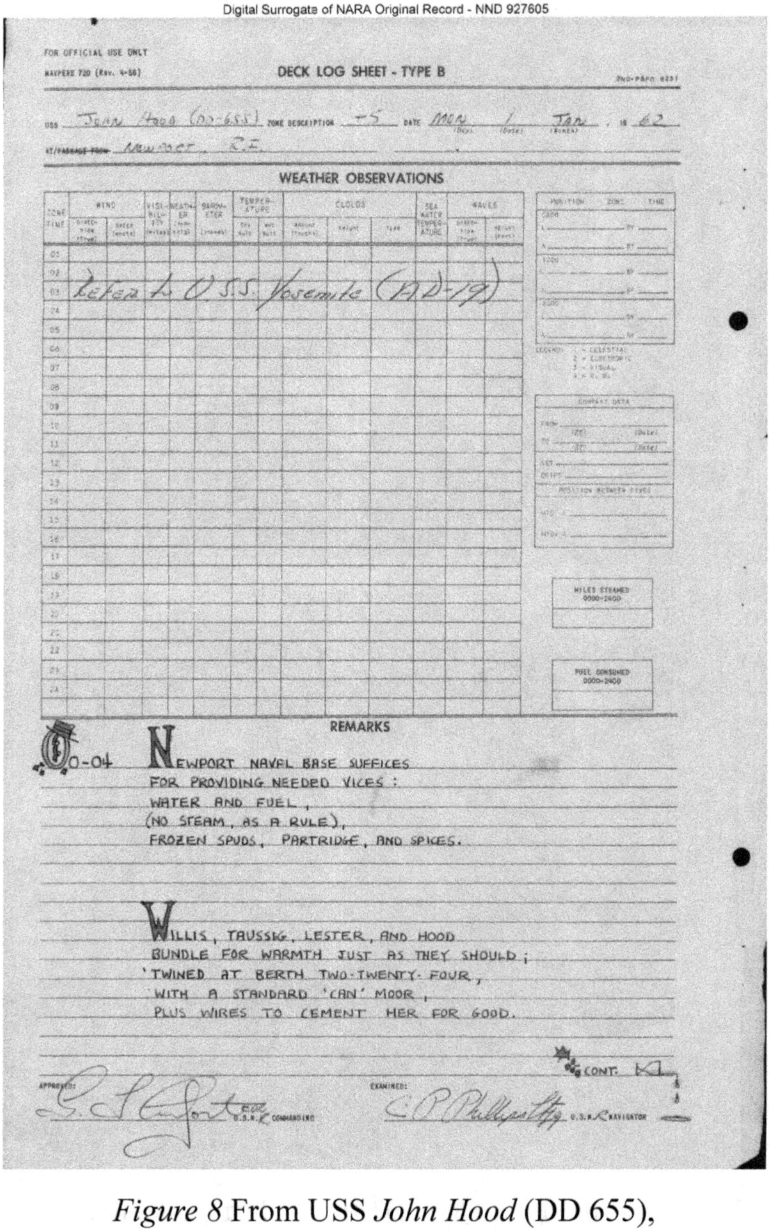
Digital Surrogate of NARA Original Record - NND 927605

FOR OFFICIAL USE ONLY

NAVPERS 720 (Rev. 4-58)

DECK LOG SHEET - TYPE B

USS John Hood (DD-655) ZONE DESCRIPTION +5 DATE MON 1 JAN 19 62

AT/PASSAGE FROM Newport, R.I.

WEATHER OBSERVATIONS

Refer to U.S.S. Yosemite (AD-19)

REMARKS

00-04 NEWPORT NAVEL BASE SUFFICES
FOR PROVIDING NEEDED VICES:
WATER AND FUEL,
(NO STEAM, AS A RULE),
FROZEN SPUDS, PARTRIDGE, AND SPICES.

WILLIS, TAUSSIG, LESTER, AND HOOD
BUNDLE FOR WARMTH JUST AS THEY SHOULD;
'TWINED AT BERTH TWO-TWENTY-FOUR,
WITH A STANDARD 'CAN' MOOR,
PLUS WIRES TO CEMENT HER FOR GOOD.

CONT.

APPROVED: U.S.N. COMMANDING

EXAMINED: U.S.N. NAVIGATOR

*Figure 8* From USS *John Hood* (DD 655), National Archives, NAID: 220042163, Container ID: 2263, January 1, 1962.

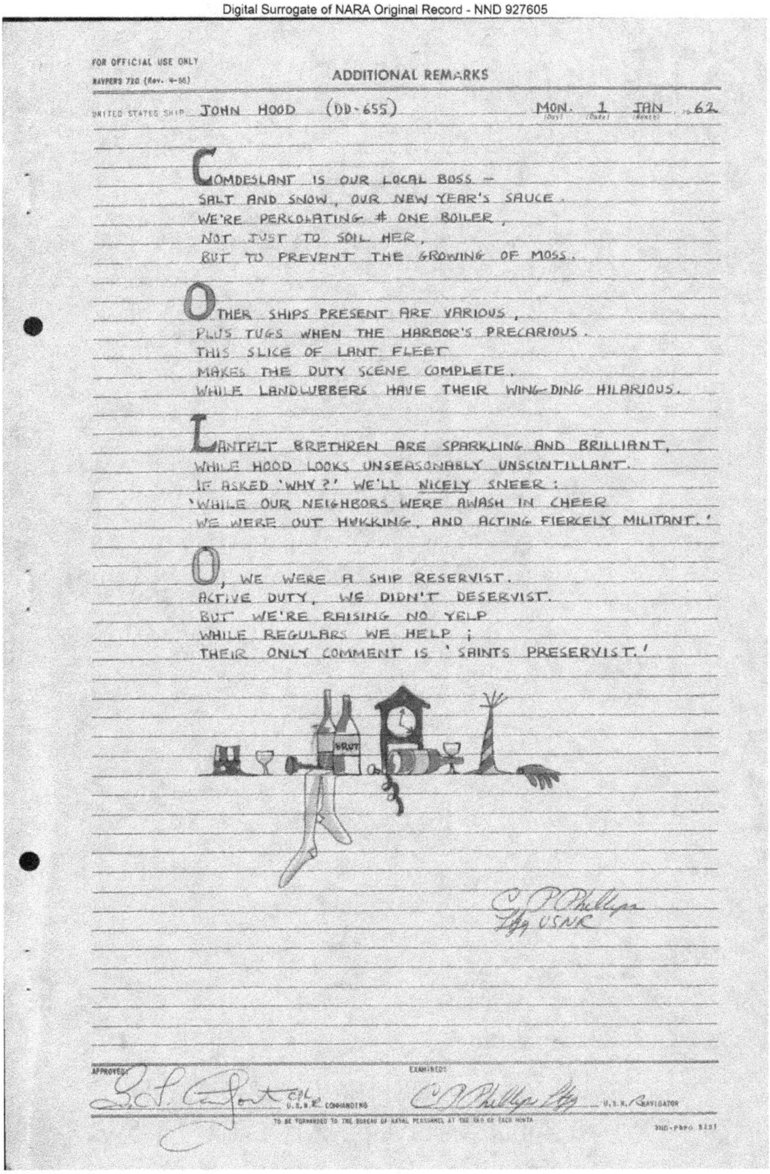

FOR OFFICIAL USE ONLY
NAVPERS 720 (Rev. 4-56)

ADDITIONAL REMARKS

UNITED STATES SHIP JOHN HOOD (DD-655) MON. 1 JAN 62

COMDESLANT IS OUR LOCAL BOSS –
SALT AND SNOW, OUR NEW YEAR'S SAUCE.
WE'RE PERCOLATING # ONE BOILER,
NOT JUST TO SOIL HER,
BUT TO PREVENT THE GROWING OF MOSS.

OTHER SHIPS PRESENT ARE VARIOUS,
PLUS TUGS WHEN THE HARBOR'S PRECARIOUS.
THIS SLICE OF LANT FLEET
MAKES THE DUTY SCENE COMPLETE,
WHILE LANDLUBBERS HAVE THEIR WING-DING HILARIOUS.

LANTFLT BRETHREN ARE SPARKLING AND BRILLIANT,
WHILE HOOD LOOKS UNSEASONABLY UNSCINTILLANT.
IF ASKED 'WHY?' WE'LL NICELY SNEER:
'WHILE OUR NEIGHBORS WERE AWASH IN CHEER
WE WERE OUT HUKKING, AND ACTING FIERCELY MILITANT.'

O, WE WERE A SHIP RESERVIST.
ACTIVE DUTY, WE DIDN'T DESERVIST.
BUT WE'RE RAISING NO YELP
WHILE REGULARS WE HELP;
THEIR ONLY COMMENT IS 'SAINTS PRESERVIST.'

C. P. Phillips
Ltjg USNR

APPROVED: [signature] CDR, U.S.N.R. COMMANDING
EXAMINED: C P Phillips Ltjg U.S.N.R. NAVIGATOR

TO BE FORWARDED TO THE BUREAU OF NAVAL PERSONNEL AT THE END OF EACH MONTH

*Figure 9* From USS *John Hood* (DD 655), National Archives, NAID: 220042163, Container ID: 2263, January 1, 1962.

*Figure 10* From USS *Willard Keith* (DD 775), National Archives, NAID: 204570459, Container ID: 5270, January 1, 1956.

# SIDELIGHT 12

## From Navy Steel to Kinetic Sculpture

James L. Seawright, Jr.[1] was born in Jackson, MS, in 1936, and spent many of his early years residing in Greenwood, MS. He earned a bachelor's degree in English at University of Mississippi in 1957. That same year, Seawright enlisted in the US Navy and received his commission as an ensign and eventually rose to the rank of lieutenant in 1962. According to one of his obituaries, he used his time as an engineering officer aboard USS *Parle* (DE 708) to learn more about mechanical and electrical devices that informed his later creation of kinetic metal sculptures. Prior to his discharge from naval service, he married Mimi Garrard, the founder of the renowned Mimi Garrard Dance Theater based in New York City. The two collaborated on innovative, visually stunning digital productions of Garrard's choreography.

Seawright spent most of his career at Princeton University as professor of visual arts in the Lewis Center for the Arts. By all accounts, his affable southern gentleman demeanor and ability to cook a killer gumbo, endeared him to students and faculty. But it was his artistic prowess that led him to worldwide fame. His work is housed in permanent collections of the Museum of Modern Art, the Metropolitan

Museum of Art, the Whitney Museum, and the Guggenheim Museum of New York. He died at the age of 85 in 2022 due to complications from Parkinson's Disease.

Here is the midwatch poem he wrote while aboard USS *Parle* in 1960. He also wrote a poem for the same ship in 1959. He clearly had a penchant for alliteration.

00 to 04

BEHOLD! Bound bravely . Six lines stilling,
The sleek ship of the sea . Starboard side to ROBERTS
Pier eleven North of her . Lieth in the lassitude
Of the New Year's midnight . Alive more than the men
Manning her dark decks. . See the silent eyes
Of the semi-sleeping sailors . Liberty's lack lamenting
Grimly, angrily guarding . Her quiet quarterdeck.
Soon, stumbling, stuporous, . Her heavy-lidded heroes
From their wassail-cups will come . Tumbling to her.
There, in the heavy redness . Of her dank holds,
Surcease seeking, . Sleep will steal them.
All around, others . Of the Eastern ocean's fleet
Many here Amphibious . (for the place hereround be LCreek,
In the Virgin Queen's lost land) . Lie adjacent, grateful,
Like we of the mighty Admiral, . COMPHIBLANT, who guards us.
Minions of him give us . Force to drive the engines
Of our sustenance; and drink . Our souls to sustain.
Steam send us they . They of the thimble-ships
Seen here severally ..

GOD grant no goblin . Or spirit stealthy
Stay the safeguards . Of our seaworthiness
Which we have set . And called YOKE ..

LET the Year bring its burden . Bountiful or brutal,
Leaders we have . And faithfully shall we follow ..

J. L. SEAWRIGHT, JR
LTJG, USNR

# CHAPTER 12

## Bards of Viet Nam

The U. S. Navy was an important player in Southeast Asia as early as 1950 and would be actively involved there until the final evacuation from Saigon in 1975. From 1964 to 1973, "2,636,000 sailors and Marines served in the Southeast Asian operational theater."[1] Navy and Coast Guard ships participated in operations in North and South Viet Nam, Laos, and Cambodia. From river patrol boats to aircraft carriers, in close combat or in coastal shelling and air strikes, the Navy was essential to the war effort. And whether in combat or in support, the daily life on Navy ships operated by the same regulations that dictated life aboard Navy ships everywhere. As a result, when the first midwatch of the year rolled around, many young officers on Navy ships in Viet Nam chose to write the deck log in verse.[2]

Of the more than ten thousand deck log poems retrieved by the *Midwatch in Verse Project*, written from the late 1940s to the mid 1970s, hundreds were written on ships involved in the Viet Nam War. An examination of those poems as compared to deck log poems written on ships elsewhere reveals both commonalities and distinct differences. The basic required information for deck logs in general are included in the Viet Nam poems, and many of them also exhibit the common concerns exhibited in deck log poems everywhere—lack of alcohol and/or women, complaints about writing a poem, unusual rhyming words, and the sailor's ever-present capacity for humor. But the nature of the war in Southeast Asia, and perhaps cultural and political developments back home, created a cluster of issues not faced in earlier conflicts and a style of poetry in the Viet Nam poems that might at times be called confessional.

### Common Elements of Viet Nam Poems and Poems Written Elsewhere

*Imitations of Classic Poems:* A rare but interesting idea that occurs in deck log poems from all times and places is the attempt to mimic the style, structure, and tone of well-known poems in the literary canon. Lt. j.g. G. J. Yablonicky aboard USS *Lofberg* (DD 759), while patrolling on Yankee Station in Tonkin Gulf in 1969, set his poem as an imitation of E. A. Poe's "The Raven." Here is the first stanza.

> Now upon this New Year dreary, as we sail with task group weary (77.5),
> Mighty Hancock (CVA 19) and sleek Decatur (DDG 31) accompany my five inch
> bore.
> Ever present this need for steaming, steaming Seventh Fleet is deeming,
> Deeming by OpSked gently reaming (3-69) to enforce a peacetime lore.
> Tis for the cause of peace this steaming, only this and nothing more.

*Absence of Alcohol and/or Women:* Alcohol had not been allowed on US Navy ships since July 1, 1914, by General

Order Number 99 (See Chapter 3), and women were, except for nurses on hospital ships, still far from a common presence onboard. A sailor's appetite for alcohol and the opposite sex may be a long-held stereotype, but New Year's deck log poems on Navy ships clearly and frequently reflect the yearning for both by young officers on duty on New Year's Eve.

Lt. j.g. J. A. Gelsomino, while on the carrier *Ticonderoga* (CVA 14) during her first full combat deployment in Viet Nam on New Year's 1966, offered the standard complaint:

> First watch of the year, holiday cruise,
> No sleep, no broads, no snooze, no booze.

*Complaints About Writing a Poem:* Though writing the deck log in verse was not required by any official regulation, many young officers who chose to do so couldn't resist poking fun at the tradition by commenting on the difficulty of meeting the challenge. One young ensign on the attack transport *George Clymer* (APA 27), after returning to San Diego from Viet Nam, used his own name and rank in the poem, pleading for mercy:

> One last word before I go
> If ever Heaven you should know
> And the Pearly Gates you to guard
> Be sure to admit this lowly bard,
>
> For he has struggled many a day
> To rhyme for you this tiresome lay
> So surely him you should not bar
> Ensign G. L. Holder, U.S.N.R.

*Unusual Rhyming Words:* The spirit of wordplay seemed to be alive and well in the Viet Nam poets. One of the common ways that playfulness was expressed appeared in the use of unusual words to create a rhyme. Lt. j.g. Thomas A. Ewing, on the guided missile frigate *King* (DLG 10), doing search and rescue in the Gulf of Tonkin in 1968, surmised,

Witless foes from any quarter
Will feel our steel a thwarter.

*Humor:* Examining deck log poems written from 1941 to 1946, the *Midwatch in Verse Project* found that, even during the darkest times in WW II, the young officers who followed the tradition of writing the New Year's deck log as a poem never lost the capacity to find humor in their circumstances.[4] The Officers of the Deck during the Viet Nam War were equally able to call on that enduring American trait. Ensign F. V McGavran, serving on USS *Henrico* (APA 45), found a humorous way to make the best of a tough situation in 1967.

New Year's Eve without drink
Is as bad as you think,
But suddenly we had a bracer;
Two months of tax
Were off of our backs
With combat pay as a chaser.

*Pride and Patriotism:* Deck log poems on Navy ships in all times and places expressed pride in their ship, their crew, and the job they were doing. The pride expressed by US sailors is often the pride in a team that "can't be beat." Lt. j.g. John R. Smutko, aboard USS *Terrell County* (LST 1157), working as part of the Danang to Chu Lai shuttle at the beginning of 1967, claimed,

A small crew we have, but they never give up
And sometimes they don't breakfast or luncheon or sup
And they don't get the medals, they don't get a cup
But there's not one in the bunch that a tiger could 'whup.'

## Distinctive Elements of Viet Nam Poems

Though Viet Nam poems exhibit many common elements with those written elsewhere, the Viet Nam poems display significant distinctives. Because of the poet's circumstances, serving on a ship actively engaged in a combat zone, his focus often veered toward the actions his ship was performing in support of the war. Though deck log poems written during WWII might refer generally to their combat role, referring, for example, to the poet's desire to blow the enemy out of the water, those poems rarely dealt directly with the combat action of the ship.[5] But poems written during the Viet Nam War frequently feature specific description of wartime activities, including interdiction, fire support, heightened security, search and rescue, and recovery.

*Interdiction:* One of the broadest and most difficult duties assigned US Navy ships in Viet Nam was interdiction of North Vietnamese attempts to resupply Viet Cong troops in the south. By 1964 it was clear that the US support role in South Viet Nam was failing.[6] After the Gulf of Tonkin incident between three North Vietnamese torpedo boats and the destroyers *Maddox* and *Turner Joy* on August 2 of that year, the Gulf of Tonkin Resolution authorized the president to use military force.

In March 1965 the Navy initiated Operation Market Time, followed by Operation Sea Dragon, Operation Sealords, and other efforts to disrupt the enemy's attempts to resupply. Navy and Coast Guard ships, along with long-range patrol planes, attempted to lock down the 1,200-mile coastline. It was impossible to stop and search every vessel. According to one estimate, over 60,000 junks and trawlers, fishing and carrying goods to markets in Danang and Saigon, cruised the offshore waters.[7] Identifying the ships secretly carrying guns, ammunition, and supplies hidden in their holds was not viable without a physical search. The stop-and-search routine became a wearisome chore.

Yet the job had to be done, and the Navy crew had to remain alert, because it was a potentially dangerous task. Lt. j.g. Robert C. Roy, on USS *Falgout* (DER 324) working in the South China Sea in 1967, described both the mission and the necessary caution.

> ComSeventhFlt did schedule this DER
> To guard against infiltration
> Originating from near or far
> To help preserve a nation .
>
> . . . .
>
> Again maneuvering to intercept
> Another possible foe
> Steering various courses, at various speeds,
> Though not for him to know.

Another ship patrolling the coast of Viet Nam on January 1, 1967, USS *Blanco County* (LST 344), was an old hand in warfare, having disembarked troops and unloaded vehicles and equipment on Omaha Beach on D-day. Only recently recommissioned in June, she joined Market Time operations in November. The Officer of the Deck on New Years, Lt. j.g. M. A. Harvey, shared the constant watchfulness required.

> Our boats are patrolling at GQ tonight
> For the New Year's cease fire is a fine time to fight
> Old Charlie—when he tries to sneak by
> With a cargo of sorts that's a mighty far cry
> From the fish and sliced shark meat he usually lets dry
> On his fantail.

When Ensign A. J. Lauinger, had the duty aboard the patrol gunboat USS *Asheville* (PG 84) on New Year's 1969, while serving with the Coastal Surveillance Force as part of Market Time, the opening lines of his poem projected a tone of confidence, bordering on cockiness.

'Tis the start of a New Year. We're out on patrol—
Watching the sampans instead of a bowl.
But we can hack it—the reason, you see,
Is that we are the men of a fighting PG.

Ensign Lauinger's attitude turns to swagger when he says,

We're off on our way, to detect what we might,
Our number one aim: to get into a fight.

*Fire Support:* From close offshore and from further out to sea, battleships, cruisers, destroyers, Coast Guard cutters and other ships provided another critical action—artillery support, helicopter gun support, and search and rescue operations. In addition, carriers launched aircraft around the clock to destroy enemy positions and disrupt over-land supply routes.

On New Year's Day of 1970, USS *Coral Sea* (CVA 43), a Midway class carrier on the fifth of her seven deployments to Viet Nam, worked the midnight-to-noon shift. Lt. j.g. L. D. Henson described their nighttime launches.

At one minute 'til one we launch our tots
Steering our ship on a course one-twenty
Churning the seas at thirty-one knots
Bringing our foe unhappiness plenty

. . . .

At zero two thirty we launch once more
Twelve planes 'gainst "Charlie," our much daunted foe
To stop all of his trucks north of I Corps.
And for freedom and peace to strike a strong blow.

USS *Davis* (DD 937), in action near the Cua Dai River south of Danang, supported a Marine unit ashore.

> The start of this new year, 1969
> Finds us hard at work along the gun line.
> As a tribute at midnight in lieu of a toast
> The Marines requested we shoot up the coast.

Not far away, in Vung Tau Harbor, USS *Samuel N. Moore* (DD 747) was also providing gun support. Lt. J. M. Smith seems to have been a fan of the popular TV series *Have Gun Will Travel*. He described the ship's mission this way.

> Our mission here
> To provide gun fire true
> To the Marine, the Army,
> And the Air Force, too.
> Have guns—will shoot.

In addition to US and South Vietnamese Navy ships, the US Coast Guard played a key role in securing the Viet Nam coastline. The Coast Guard contingent during the war consisted of 56 separate combat ships and approximately 8,000 men.[8] On New Year's Eve 1969, the Coast Guard cutter patrol boat *Point Hudson* (WPB 82332) was providing close-range coastal fire support with her 50-caliber machine guns and 81-mm mortar, and Lt. j.g. Robert H. Anschuetz shared the details in his deck log poem.

> As our first act of the New Year, we commence
> To fire Hotel and India for Muddy Rifle 14 and his gents.
> 10 minutes is all that is needed
> To show "Charlie" he will surely be defeated.
>
> 900 rounds from our 50 caliber machine gun
> And 8 more illumination from mount eighty-one
> Fire power is no great worry
> Because we put the ammo out in a hurry.

We know we put "Charlie" on the run
And we are glad to announce, "friendly casualties, none."
And so to Uncle Ho and his gang
We hope we started your New Year off with a great
big bang.

*The River War:* One arena of combat in Viet Nam was new to the US military and required the creation of a new kind of force - the Brown Water Navy. When the Vung Ro incident in March 1965 showed the Vietnamese Navy to be ineffective, the US Navy moved in, beginning its river war.[9] The riverine force became an essential element of the conflict in the vast network of river systems in the Mekong Delta and elsewhere, and when the Viet Cong attacked numerous sites in the Tet Offensive of 1968, the river force was credited by some with saving the Delta.[10] They wore black berets, often fought in close combat, and exhibited great pride in their identity.

Ensign E. S. Gilmore Jr., as Officer of the Deck aboard USS *Benewah* (APB 35) in 1970, while anchored in the Mekong River near An Long, exhibited a mischievous attitude toward his superior officers and deep pride in the work being done by the riverine force.

Something to add 'bout our old sea coot,
Solomon's the name and he's SOPA to boot.
He leads a proud ship in this foreign land
In a terrible war that few understand.
The sailors who've fought and come back alive
Will look back on the brown-water navy with pride.
The new year begins on this warm moonlit morn
Mid the thunder of cannon and shells it is born.

On New Year's Day 1968, Ensign W. R. Hughes, aboard USS *Garrett County* (LST 786), anchored in the Co Chien River near Cu Lao Ray, expressed not only pride in the River

Patrol Force but a typical American confidence expressed as bravado.

> Charlie's around here, that you can bet
> And for that reason, condition three is set.
> The helos on deck are poised for some action
> For the boats on patrol, they're ready reaction.
> The little green boats tied alongside
> Ride with the currents and turn with the tide.
> The River Patrol Forces, that's what we are,
> Our homes are all stateside, so near but so far.
> When this war's over, and back there we head
> Viet Nam will resound with, "Charlie is dead."

It wasn't only combat ships that made the riverine forces so formidable. Support ships also worked in the rivers and showed pride in their work. The light repair ship *Sphinx* (ARL 24), operating in the Song Vam Co River, was doing its part to keep the combat ships in action. Chief Petty Officer R. M. White wrote in his deck log poem,

> This green ship's hard working and if any should doubt us
> The river assault groups wouldn't last long without us.
> SOPA is CO of Sphinx ARL twenty-four,
> And we're doing our part to win this DAMN WAR.

*Vulnerability in the River War:* Navy ships, when anchored in one of the many rivers that snaked through the Mekong Delta and other areas, were especially vulnerable to attack. One of the most feared threats to Navy ships consisted of Viet Cong sappers, with a limpet mine strapped to their bellies, swimming to a US vessel and planting the magnetic mine. When anchored, ships were on high alert and used various tactics to avoid tragedy.

When the light cargo ship *Mark* (AKL 12) was docked at Vinh Long in the Co Chien River in 1968, Lt. j.g. W. S. Boyd expressed the tension felt aboard the ship.

> A sentry is posted at both bow and stern.
> There and amidships, strong flood lights burn.
> The lights are arranged to show water lines.
> To discourage the VC from planting mines.

On New Year's Day 1969, Lt. j.g. G. C. Chapman was assigned the duty aboard USS *Colleton* (APB 36) while the ship was anchored in six fathoms of water on the Song Ham Luong River in the Mekong Delta near Dai Dien. A year after the Tet Offensive of January 1968, Chapman was alert to the possibility of another Tet occurring. The rivers of the delta were tidal, and when the ocean tides were moving in or out, the strong current discouraged sappers. But in that interval between tides, there was no current, and sappers could easily swim to the ship. During that time, ships spun their screws to create their own potentially lethal current. Chapman said,

> For defense condition two is set
> Should the VC attempt another TET.
> To fire on Colleton would be a mistake
> For our guns are manned and ready to rake.
>
> . . . .
>
> At zero zero thirty-five
> The ship's main engines came alive.
> The screws were spun as a mine deterrent,
> Since swimmers could attack when there is no current.

Another line of protection for ships anchored in harbors or rivers was the use of converted LCVP's from WW II as patrol boats, whose main job was to keep an eye out for swimmers. Ensign D. D. Michalske, aboard USS *Hunterdon County* (LST 838), anchored in the Bassac River in 1969, mentioned the old landing craft.

> The LCVP on security patrol
> With protection from sappers the primary goal.
> Condition three (modified) watch has been set.
> Prepared for the possible Viet Cong Tet.

One of the more aggressive attempts to discourage sappers was described by Lt. j.g. W. H. Kugel, manning the quarterdeck in 1970 at Vung Tau, "watching bright flares, hoping rockets won't hit." He revealed the tension felt by the crew, saying,

> Fourteen fathoms of water, sand and mud bottom
> We watch for enemy swimmers and movement of flotsam.
>
> . . . .
>
> Of grave concern is watertight integrity,
> And measures we've taken for special security.
> We adapt our eyes and use visual aids,
> And just to make sure drop percussion grenades.

*Recovery Operations:* The concern over mines was well placed. When despite all precautions a sapper did sneak through and attached a mine to a Navy ship, the job turned to recovery operations when possible. The effort could be difficult and discouraging.

On Thanksgiving night in 1967 at Ben Tre, the repair, berthing, and messing barge YRBM 16 was blown up by a mine, resulting in five sailors killed and two wounded. A little over a month later, the fleet ocean tug USS *Abnaki* (ATF 96) towed the damaged ship to Sasebo, Japan, from Dong Tam, Republic of Viet Nam. On board the tug Signalman First Class D. L. Ellis sounded weary when he wrote,

Riding on the brake at 1500 feet, YRBM 16 in tow
Since Christmas, we've been underway,
Up and down the river, scraping bottom half the way.

And to add to his discomfort,

In 15 foot seas we're tossed to and fro,
At 25 knots the monsoon winds blow.

On December 22, 1968, the Shell tanker SS *Helisoma* was mined in Nha Trang Harbor by a suspected Viet Cong swimmer. The mine blew a hole 10 feet 8 inches by 11 feet 7 inches, 16 feet from the bow and 3.5 feet below the waterline, puncturing cargo tanks filled with jet fuel.[11] On December 26 USS *Safeguard* (ARS 25) arrived to begin repair work. *Helisoma* was 169 meters long with a beam of 21 meters. *Safeguard* was 70 meters long with a beam of 12 meters. Saving the *Helisoma* was a big job. On New Year's Eve, the OOD on *Safeguard*, Lt. j.g. W. S. Schaill, wrote that the New Years bells had sounded, and the flares had fired to celebrate,

Yet the job remains.
Still the Helisoma,
Shell tanker,
Sits moored to our port side—
Sinking slowly,
Sinking slowly.

Our doubled lines hold us in place;
But they can't keep her afloat.

Only our pumps,
With our ship's power,
Keep us both alive.

Despite the somber tone of the poem, the hole in *Helisoma*'s bow was patched, the jet fuel transferred to other ships, and after a "great New Year's Eve celebration," the job was completed. On January 2, Helisoma sailed for Singapore at 14:30 hours for permanent repairs. [12]

On New Year's Eve of 1970, Lt. Stephen D. Chubb on USS *Current* (ARS 22) described the attempt to raise an LCM 6 landing craft from the bottom of Danang Harbor. An armed patrol boat watched over them as they worked. Lt. Chubb closed the poem with the last two lines of Robert Frost's poem "Stopping by Woods on a Snowy Evening."

> Mike-6 below we've sworn to raise
> Two lines attached in two full days
> Soon's the time that from the deep
> Up she'll come with loud Hoo-rays
>
> An armed patrol a watch does keep
> While we prepare to cheat the deep
> For we have promises to keep
> And miles to go before we sleep.

*Human Impact:* Some of the poems written in Viet Nam revealed not only the dangers of combat but also the emotional toll the war took on the sailors involved. Lt. H. W. Clark Jr, writing on New Year's Day 1968 shortly after USS *Henry W. Tucker* (DD 875) returned from Viet Nam, recalled search and recovery missions for downed pilots and the recovery of survivors of the tragic fire on the carrier *Forrestal* (CVA 59) the previous year, which killed 134 sailors and injured 161.

> Returned from the war, completed our mission
> We spent long hours of duty and much ammunition.
> We searched for pilots off Haiphong,
> And for Forrestal survivors on that day sad and long.

USS *Page County* (LST 1076) had also just returned from Viet Nam in January 1969 when Lt. j.g. A. A. Owens wrote in his deck log poem of memories that are deeply set but not easily shared.

We pulled into San Diego December twenty four
From a long tough haul in that South China war
This LST so hearty and bold
Holds many a story that can never be told.

. . . .

This past year was rough, but the men saw it through
When things really got tough, they knew what to do
We've fought at Danang, Chu Lai and Cua Viet
These places no man will soon forget.

There were, however, officers whose service in the war encouraged a different perspective from those serving on combat ships—the officers serving aboard naval hospital ships. Lt. J. A. Sargent, on the USS *Repose* (AH 16) cruising from Danang to Chu Lai in January 1968, expressed the more humanitarian purpose of their work in his midwatch poem.

With flood-lighted sides, red crosses on white
The ship sails purposefully through the night
Providing a healing hand for those who fight
Medical skills, not guns, are our might.

Later in the poem Lt. Sargent revealed his typical American humor in the middle of a difficult situation.

The nurses, those creatures of perfumed delight
Can sleep soundly throughout this long night
Knowing that condition YOKE has been set right
And all of the doctors are probably tight.

*Trouble at Home:* America's wars, from the Revolution to Operation Enduring Freedom, have had their detractors back home. But the social upheavals in the U.S. during the

Viet Nam War created the largest anti-war movement in American history. Much has been written and said about the lack of welcome and lack of support for veterans of the war when they returned home. The deck log poets in Viet Nam were aware of the problem, and some addressed it in their poems.

Lt. Robert M. De Mais, the OOD on USS *Wright* (CC 2) in 1966, summarized in two potent lines the tension between the conflict overseas and the struggle back home. Beginning on Christmas Day in 1963 in Champaign, Illinois, Vietnam War protestors had begun publicly burning their draft cards. And on October 15, 1965, just two-and-a-half months before DeMais wrote his deck log, the first prosecution under the Draft Card Mutilation Act of 1965 occurred to widespread media attention.

> Viet Nam's war is raging, there is great loss in this fight
> While standing on the White House lawn draft cards
> are burning bright.

Lt. j.g. David L. Sweet, aboard USS *Westchester County* (LST 1167) also in1966, offers a typical lament.

> We carried stores, and many men;
> Some perished in the fight:
> But we, and they, firmly believe
> What's done is for the right.
> Most folks at home support us, but---
> Some don't. A saddening few.
> Our silent answer shame their cry—
> We're struggling for them too.

The deck log poem from USS *Coontz* (DLG 9) in 1968, written by Lt. R. J. Pozzi, expresses several of the typical themes of deck log poems but also includes an extended reflection on the lack of support back home. Lt. Pozzi's poem is creative and witty. *Coontz* had just returned from duty in Viet Nam. Pozzi demonstrates his hand at humor, taking the opportunity to comment ironically on the Navy's appetite for instruction and documentation.

> One of the problems that our great Navy faces
> Is the race to fill paper with letters and spaces.
> The volume of instructions is fast growing so large
> That soon we will have to tow it behind us, in a barge.

But the humorous tone of the deck log shifts to a somber sense of bewilderment when Pozzi turns to the topic of the war protests going on back home.

> The war is unpopular with a few at home, no doubt,
> But where principles are involved, there is no easy out.
> A paradox, it seems, to us on this end
> To see the hippies take license with the freedoms we defend.
> Our ship has been fortunate no hits to receive
> But we've been close enough the horrors of war to perceive.
> For those who claim greater perception from farther away—
> How do they overlook their own freedom every day.
>
> A utopia this world is quite obviously not.
> But the promise of peace and improvement comes not from pot.
> 'Tis well to question, to seek the truth, to doubt,
> But not to throw in the towel before the end of the bout.

## Conclusion

Combat is an intense, emotional, unforgettable experience. But, although sailors in combat who chose to follow the old Navy tradition of writing the first deck log of the year in verse shared many of the same foibles and frustrations of ship life and midwatch duty that sailors everywhere experienced, the young Officers of the Deck on Navy ships directly involved in the war in Viet Nam found it difficult not to bring that experience to the foreground of their poems. They were young. They were American. And their humanity overcame the difficulty of their circumstances when they were given the opportunity to write a different kind of deck log. Their poems are now an obscure historical record of a particular time and a particular place in 20$^{th}$-century American experience. It is good that we remember.

# SIDELIGHT 13

## A Navy Man for Life

If you are sad and lonely,
If you are feeling blue.
Come aboard the SHRIKE tonight;
See others just like you.
It's New Year's Eve, it's cold and bitter;
The gob on board is a pitiful critter.
His shipmates ashore may drink to the foam,
Drink to sweethearts, folks, and home.
But the saddest of all in the U.S. Fleet,
(As he warms his hands and stamps his feet)
Is the soul with the midwatch who finds it's time
To write (ugh) poetry, verse, and rhyme. . .

Who is the sailor who writes these blues?
Why, no one else but Wayne P. Hughes . . .

As Lt. j.g. Wayne P. Hughes, Jr. penned these lines in his midwatch log on Sunday, January 1, 1956, aboard the minesweeper USS *Shrike* (MSC 201), it is unlikely he could have imagined he would one day be described as a naval icon. A 1952 graduate of the U.S, Naval Academy, in a few short years he would join the Academy staff, where he would play an important role in restructuring the curriculum to make it reflective of college education of the era, and in 1964 he would complete a master's degree in the Naval Postgraduate School (NPS).[1] His seagoing service included both Korea and Vietnam, and he commanded ships, including the minesweeper USS *Hummingbird* (AMS 192) and the destroyer USS *Morton* (DD 948).

Hughes eventually returned to NPS, where he became a beloved, highly respected teacher and scholar, described by NPS president and retired Vice Admiral Ann Rondeau as a "national treasure" and "a giant among naval strategists."[2] His writings, including several editions of the book *Fleet Tactics*, and many articles, became key resources for students of naval strategy and tactics. Writing in the *Naval War College Review*, Hughes[3] made clear his view of the critical role of tactics in naval preparedness:

> The reason that a discussion of tactics is appropriate when discussing contemporary strategy is that strategy must rest on the rock of combat capability. One builds decisions from the bottom up: tactics affect the efficacy of forces; the correlation of forces reveals what strategy our forces can support, and a supportable military strategy governs national aims and ambitions.

Hughes went on to explain how the tactical approach differed from the top-down approach often taken for granted and employed by government officials in setting national goals and strategy, with the implicit assumption that tactical forces will be able to execute them.

His work influenced U.S. Navy warfare strategy, and he was well known to numerous Chiefs of Naval Operations. Hughes was a gifted teacher, and many of his students went on to become captains and admirals (Vencill, 2019). He retired with the rank of Captain but would eventually spend forty years as an educator at the NPS, where he became

Dean Emeritus of the Graduate School of Operational and Information Sciences.

In 2024 Hughes was inducted posthumously into the Surface Navy Association Hall of Fame, which recognized him for his pivotal role in shaping naval force structure and doctrine. According to his colleague Captain Jeff Kline, Hughes's work could be captured in a phrase: "Fire effectively first (Vencill, 2019)." A music lover, he was a fan of both symphony and Dixieland jazz, but his favorite song remained The Navy Hymn. This generous pipe-smoking grandfather, mentor, and scholar died in 2019 after devoting a lifetime to the Navy. He travelled a long way from that cold winter night in 1956, when he ended his midwatch poem with these words:

> The battle of Pen and Poem is said,
> It's time all sailors were in bed,
> So now to you my final licks:
> HAPPY NEW YEAR '56..

# CHAPTER 13

## Postscripts

When young officers wrote a New Year's deck log in verse, they sometimes had afterthoughts, details they felt needed to be noted, occurring to them after the log poem was written. Approximately one in every hundred deck logs include a postscript, and the content varies widely. Often, a postscript simply catalogs a common ship activity, such as a change of course and speed, refueling, embarking, or performing drills. Occasionally, the Officer of the Deck writes a short poem, apparently just for fun, and then adds a standard prose deck log as a postscript.

The great majority of postscripts to deck log poems are simply a brief note written in prose, not verse. But a few of the young officers writing the deck log in verse seemed to enjoy their bardic duty so much that they continued the rhyme in their postscript. Lt. j.g. M. G. Meyers wrote the following note while on *Newport News* (CA 148) on New Year 1957.

> P.S. Yoke is set on second deck,
> And either above or below,
> And XRAY is the other –
> Or vice-versa, I never did know

P.P.S. Ships present are:
TI., WIS., and FDR.

Lt. C. L. Fuld, while serving the first midwatch of 1961 on *Hopewell* (DD 681), wrote,

POSTSCRIPT—
Congress may an officer make
But they are not all poets.
Though we strive to do our best
Our poetry doesn't show it.

When the news a young officer shared in his postscript went beyond typical ship activity, the postscript was sometimes more interesting than the poem itself. Sailors often felt the stress of extended time at sea. When they went ashore for liberty on New Year's Eve, letting off steam could be a dangerous activity. (Names of injured sailors redacted in following examples.)

### Fights on Shore

Sailors on liberty have long been known for enjoying their brief freedom. Their joy is often enhanced by the consumption of alcohol. Overindulgence often led to trouble— disagreements, which led to fights, which led to sailors being returned to the ship by shore patrol in less than ship shape. A sailor serving on USS *Kenneth Whiting* (AV 14), suffered a laceration to his right ear when he was hit by a bottle. The OOD, J. C. Anderson, chief warrant officer, notes in his postscript that the wound was not the result of the sailor's own misconduct and that the duty corpsman administered treatment, after which the injured man was placed on the "binnacle list," a term derived from the old Navy practice of pinning a sick list to the binnacle (the compass stand on the bridge) so the captain would be informed.

A machinist mate serving on USS *Boston* (CAG 1) was hit on the head with a bottle while on liberty in Genoa, Italy, by an unknown person, resulting in a scalp abrasion and laceration to the right cheek. The wounds were cleaned and dressed, and he was returned to duty. Ensign E. P. Allen, the OOD on *Charles J. Badger* (DD 657) harbored at Cannes, had a bit of fun writing his postscript in verse while reporting the injury.

> At 0045 on this rainy night
> A yeoman third with cut lip returned
> Said he got hit while settling a fight
> And he for the duty corpsman yearned.
>
> --------, Gerald (No middle name)
> [most of serial number deleted] eight-eight
> His lip quite lame
> Had to for the corpsman wait.
>
> The corpsman came and sewed his face
> One suture and bandage did the trick
> -------- for his bunk did race
> And sound asleep he fell real quick.

A sailor from USS *Fort Snelling* (LSD 30), docked at Genoa, returned aboard under his own power "after being hit and kicked in the face by an unknown assailant from USS *Monrovia*." He had apparently suffered a serious laceration below his right eye. The medical staff of USS *Telfair* treated the wound and had to give him 50 mg. of Librium to calm him down. A sailor from USS *Orion* (AS 18), docked at the Naval Station, Norfolk, Virginia, returned aboard complaining that "the marine sentry at Destroyer/Submarine Piers gate struck him on left side of face." No details of what prompted the marine to strike him are given in the postscript. All that can be gleaned from the brief report is that the Marine was probably right-handed, and the sailor undoubtedly wished he had not done whatever he did.

### Accidents on Shore

In addition to fighting while on liberty, accidents of various kinds plagued celebrating sailors. A Naval aviator was handling black powder around an open flame when the powder exploded, causing a first-degree burn on his right hand. He was treated by the medical officer and returned to duty. At 0210, the OOD at a Naval Air Station reported that a Fireman Apprentice had fractured his jaw in an automobile accident and had been admitted to the hospital. And Lt. Leighton B. Brown, the OOD on USS *Little Rock* (CLG 4), ended his midwatch log with a postscript in rhyme describing a sailor with evident anger management problems.

What's this here? An injury report I have received.

> A man slugged a door and broke two fingers I believe.
> ---------- is his name, ----- ----, an FN it says here.
> I think he had imbibed a little too much New Year's cheer.
> On the street between the Hideaway and fleet landing
> He says he struck a door near where he was standing.
> The X-ray showed fractures; an ice pack was applied.
> And now with a splint, taking Darvon in sick bay he lies.

According to a deck log postscript from *Franklin D. Roosevelt* (CVA 42) while docked at Cannes, France, a sailor on liberty stubbed his toe on an unknown object, resulting in the traumatic removal of his left toenail. The injury was treated by the medical officer, and the sailor was returned to duty.

## Accidents on Board

Though liberty and alcohol were a risky combination for sailors on New Year's Eve, staying on ship as part of the duty crew also carried its own risks. Injuries and emergencies were common occurrences and often of a droll nature. According to one postscript, at 0225 on New Year's Eve 1956, a sailor aboard USS *Salisbury Sound* (AV 13) was placed in the sick bay ward for observation by the medical officer after receiving a one-inch laceration on his chin when he fell down a ladder from 02 deck to 01 deck. In another postscript from USS *Fiske* (DDR 842), a machinist mate was reported taken to the base dispensary for treatment after suffering a laceration above his left eye when his foot slipped off a rung of the stool he sat on and he fell off. In the words of Lt. j.g. D. L. Hanson, at 0300 on USS *Carpenter* (DDE 825) a commissary seaman "fell off the dock and landed on the camel [an external flotation device]. Alongside starboard aft. Ambulance called and man treated for shock on camel. 0310 ambulance arrived and patient removed from camel by ambulance personnel. 0315 ambulance departed for Tripler Army Hospital with [the injured man]."

## Emergencies on Board

One of the most common emergencies that occurred on ships was fire. The combination of close spaces, abundant fuel, operating machinery, and a large percentage of men who smoked offered ample opportunity for ignition. On USS *Macon* (CA 132) on New Year's 1960 at 0047, a postscript to the log reported a fire in the number three fireroom. The fire involved two swabs in the upper level of the fireroom and was extinguished at 0049. Apparently, a lighted cigarette ignited the fire, and no damage was reported. That same night a postscript to the log on USS *Tidewater* (AD

31), reported at 0310 the security watch discovered rags burning in the cobbler shop. The fire was extinguished, and the cobbler shop was secured.

Though fire was a common emergency onboard ship, other serious incidents were reported in log postscripts on occasion. At sea on New Year's Eve 1959, USS *Yorktown* (CV 10), reported a man overboard. Lt. j.g. E. T. Boris, OOD aboard *John S. McCain* (DL 3), one of the screening ships, received the report at 0002, and *McCain* joined the search. But at 0031 Boris noted in a postscript that all personnel on *Yorktown* were present and accounted for. In this case, it was a false alarm. Sometimes the emergencies reported were simply medical. On January 1, 1956, Lt. j.g. W. J. Lane reported in a postscript from USS *Stockham* (DD 683), "0130 [sailor] left the ship in company with the HMC from the REMEY (DD 688) for the GRAND CANYON, with a 'possible appendicitis'."

## Disciplinary Issues on Shore

Frequently, what the OOD noted in his postscript involved disciplinary or legal matters concerning members of the ship's crew. Sailors on New Year's Eve liberty consuming too much alcohol sometimes made poor decisions and were called to account by either military or civil authorities. By far the most common offence listed in the deck logs was being away without leave for a few hours or occasionally a matter of days. In most of these cases, the OOD noted the offence perfunctorily. However, on USS *Coral Sea* (CVA 43) in 1963, the OOD was 2nd Lt. J. A. Shepherd of the US Marines, and he was clearly not happy about a sailor returning from unauthorized absence. His postscript read:

> I learned that at 2355 --------- returned from over the hill,
> And I in disgust spilled my ink and returned to my quill.

Often a sailor's offence on shore had to do with the misuse of an automobile. A postscript to the log of USS *Roanoke* (CL 145) in 1957 revealed that a sailor was arrested for breaking into an automobile. He resisted being taken into custody and was detained for the break-in, being out of uniform, and "breaking arrest." The commanding officer of *Roanoke* ordered him confined "for safe keeping." The OOD from *Long Beach* (CGN 9) in 1962 reported in his postscript that a sailor attempted to break into a civilian automobile, again out of uniform, and in this case "violently resisted when Civilian Police and the shore patrol apprehended him." The young man refused to be treated at a civilian hospital and so was held by the senior medical officer of the U. S. Naval Station until later that day. Civil charges were pending. In one case reported in a postscript on USS *Radford* (DDE 446), a sailor was returned to the ship, again in civilian clothing and in custody of the shore patrol, and charged with "Wrongful appropriation of a vehicle," an official sounding way of saying he stole a car. In another incident, according to a postscript, the Norfolk Shore Patrol delivered a sailor to USS *Laffey* (DD 724) after receiving him from the Norfolk police, who had arrested him for reckless driving and hit and run. He was placed on restriction and scheduled to appear in traffic court in two days' time.

The long arm of the shore patrol took sailors into custody for a variety of other offences committed while on liberty, and the incidents were duly recorded in deck log postscripts. At Ford Island, a Coxwain, who had accepted custody of two sailors from the shore patrol, returned the two men to USS *Hamul* (AD 20) charged with "Breach of the peace (Fire works violation)." A sailor from *Kittyhawk* (CVA 63), docked in Yokosuka, Japan, got a free ride back to the ship courtesy of the Yokosuka shore patrol when they found him asleep in a bar.

## Disciplinary Issues on Ship

But it wasn't only the sailors on liberty who occasionally made poor decisions. It seems that some of the men confined to ship for duty on New Year's Eve just couldn't bring themselves to forsake all the celebrations. Since alcohol was forbidden onboard ship, the attempt to celebrate covertly carried significant risk. Simply possessing alcohol was cause for discipline. When consuming alcohol led to trouble on board ship, the punishment could be swift. On USS *Amphion* (AR 13) in 1964, two sailors drank their stash of alcohol and created a disturbance. The postscript to the deck log notes that the CDO immediately ordered them "confined to the brig for safe-keeping." An even more serious offense involving alcohol on board was being drunk on duty. A sailor aboard USS *Huse* (DE 145) in 1956, while the ship was on antisubmarine operations, was caught, according to the log postscript, drunk on duty, "disorderly and incapacitated." He was arrested and incarcerated in the Naval Station Brig at Newport, Rhode Island.

Not all offences committed aboard ship and recorded in midwatch postscripts resulted from bringing alcohol on board. Sometimes men on a ship just did stupid things. On New Year's Eve 1951, two fire controlmen on USS *Adirondack* (AGC 15) were caught up in unusual hijinks. The OOD ended his midwatch postscript by reporting the incident in rhyme, using his own name for the final rhyme.

> . . . . . . . . and . . . . . . . ., fire controlmen third
> Were placed on report for a loud banging we heard
> It turned out to be the starboard saluting gun
> Without proper authority they were having fun
> The C.O.'s order made them prisoners at large
> And now they will muster with the MAA sarge
> My relief is coming, I hear him, I think
> Respectfully yours, Chief Carpenter Fink.

There could, of course, be other serious concerns. The postscript to the first watch of 1958 aboard USS *Macon* (CA 132) records two such incidents. The Commanding Officer, at 0220, ordered one sailor put on report for dereliction of duty. No details were offered in the postscript. But another sailor was placed on report "accused of being disorderly in quarters, of communicating a threat, of assaulting a petty officer, of being disrespectful in language toward a petty officer and failure to obey a lawful order."

**Miscellaneous Events**

In addition to the litany of injuries, emergencies, and libelous actions, deck log postscripts mention a miscellaneous grab bag of events on a ship. Lt. j.g. R. L. Helmreich, serving as OOD on the destroyer *Decatur* (DD 936) in 1963, noted that the Duty Driver reported that the ship's vehicle, the *Travelette*, had been stolen. The Boston Police Department and the Armed Forces Police had been notified. In 1959, the OOD on the ice breaker *Staten Island* (AGB 5) noted in his postscript that at 0324 the ship began breaking a channel to McMurdo Station, which was established just four years earlier in 1955 but eventually became the largest US Antarctic station. On New Year's Eve 1960, the postscript of Ensign J. T. McKissen, aboard *Samuel N. Moore* (DD 747), noted simply at 0001, with tongue in cheek, "All New Years resolutions are, as of now, in effect." And that same night Lt. J. O. Rogers, on auxiliary submarine *Baya* (AGSS 318), ended his log with the following parenthetical postscript in the form of a limerick:

> (Editor's Note:
> On Baya we have a navigator named Bacchus
> If we don't write our logs he'll "in-hack" us
> So to stay off the hook
> I write poems in this book
> Although it makes me feel like a jack-ass.)

In the long history of midwatch deck log poems, the postscript seems to have served not only as an opportunity to insert information not included in the poem itself but also as an opportunity for the young officers writing the log to display their sense of humor, despite being stuck on duty while their friends were celebrating. Capt. B. A. Adams, USMC, on USS *Macon* (CA 132) in 1958, penned a long poem, then added a postscript that mixed poetry, prose, and more poetry to wrap up his log. The postscript begins,

> P.S. It's criminal, it's criminal, a horrible shame
> A sailor's good name to bard or defame,
> Yet duty is duty and reports I must add
> Not written in rhyme – I'm a terrible cad.

These poetical lines are followed by a non-poetical prose description of the two sailors mentioned earlier accused of dereliction in the performance of duty and other offences. The log then ends, as does this chapter, with these lines from the postscript.

> If you've read all of my song, my thanks I extend
> And repeat one more time this is really the end.

# SIDELIGHT 14

## This, That, and the Otter: Playing with Language on New Year's Eve

It is no surprise that Navy midwatch poets, disappointed at having duty on New Year's Eve and bored by the long mid-to-four watch, found numerous ways to help pass the time. One option was to make the midwatch poem interesting by playing with the English language. Wild misspellings, made-up words, and tongue-in-cheek imitation of other languages were all fair game.

On New Year's 1962, Ensign J. P. Hughes manned the helm of USS *Paul Revere* (APA 248), which had launched in 1953 as *Diamond Mariner* in the merchant fleet with MARAD. Hughes's intentional misspellings imitate middle English. He closes the poem with

> We knowe we're missynge aul the festyve funne
> But we'll feele bettyre neathe the morrow's sunne
> And then twill be our tourne to shoutte and cheere
> And wishe our muddled maytes "Happee Newe Yeare.

Lt. j.g. D. C. Kapple also imitated Middle English but also echoed the opening lines of Geoffrey Chaucer's *Canterbury Tales*.

> Wenne Newe Yeare hath his Course begune,
> Yon Norris underwaye doth runne.

Lt. j.g. E. R. Shannon, officer of the deck on USS *Norris* (DDE 859) in 1960, uses the made-up word "pentametize" to describe poeticizing the log and "natch" for "naturally. He also uses the letter S shaped like a modern italicized *f,* to refer to the ship's captain as "boss." This use of s/*f*, known as the medial S or the long S, was used from the 8th century up to the 19th century in English writing. Shannon's application of it to the word "boss" visually creates possible double meanings. The word "boff," modern spelling, means a hearty laugh, a highly successful thing, or in vulgar slang, sexual intercourse. (Shannon apparently felt the need to add a parenthetical explanation to his usage.)

> To pentametize the goings on
> And natch the goings off
> Of each and every single thing
> Required by the bo*ff* (Olde Engli *f*).

Lt. j.g. T. J. Krilowiez, aboard USS *Trigger* (SS 564) in 1961 wrote his log in a cheesy stage German.

> Mine shep ist side port mit Darter, da VEV SS 578.
> Vee ist ot tree peer, und vee ist tide strait.
> Mit Darter da side port mit Gilmore,
> Vee doan hov nodding but vait.

Lt. j.g. J. R. Griffith, on the bridge of USS *Myles C. Fox* (DDR 829) in 1963, wrote his log in cheesy stage Italian, focusing on only one small usage shift, the stereotypical frequent insertion of an "a" sound where it doesn't belong.

The Fox she's - a moored in berth
Fifty-one.
The year she's - a through; in fact,
She's - a done.
Unhappily, in Napoli, the mid - a
Watch we got,
And merriment and gaiety, this we
Have - a not.

Playing with language, using their wit to create humor in the midnight watch, even though the audience was very small, was just one of the ways young officers gave the log a human touch.

# EPILOGUE

Through five years of research and the writing of two books—*Midwatch in Verse* and *The Watch Stands as Written*—we have had many highs and lows. But perhaps the highest of the highs has been the relationships developed with the surviving family members of young officers who wrote the collected midwatch poems and a few of the poets themselves. *Midwatch in Verse Project* (https://midwatch-in-verse.com/) began as curiosity on the part of author David Johnson who stumbled onto the poems while researching his father-in-law's ship, USS *Patterson* (DD 392), from WWII. But when Dave had the idea of sending a sailor's poem to surviving family members, a whole new endeavor opened and has proven to be highly rewarding and enjoyable. Surprisingly, none of the families had any knowledge of their father or grandfather or uncle having written the poems. The initial responses of surviving family members and poets to receiving a poem "out of the blue," ranging from shock and surprise to joy and excitement have been touching and inspiring for the authors.

The common thread in all the responses has been thankfulness. The unexpected gift of the poem, appearing from the distant past, brought to life the family members' poet relative, and they often learned more from the information we shared with them than the poet was ever willing to share himself. The sincerity and emotional depth of the responses became quickly evident. Brad, a poet's son, described his reaction to receiving the poem: "It's been 2 months since I opened your letter and sat there slack-jawed at what I was reading." Brad had shared the poem with all his siblings. Mary Ann, the daughter of one poet, said, "Thank you again for this information. Hearing anything about my dear dad is always a blessing that fills my heart with joy!" And Steve, the son of another poet, said, "I was very excited to receive the poem my father wrote on Jan. 1, 1945. I had no idea he had written it, since he never talked to me about his wartime experiences. I called my mother, his wife, who is alive and well at age 101. . . . Thank you, Mr. Johnson, for opening so many doors my Dad kept resolutely shut during his life with us."

Others responded in a similar way. Alice, the daughter of one of the poets, wrote, "Thank you for sending our father's poem to my brother George. It was a window into his time in the Navy. Our parents who were both in the Navy rarely talked about the war. We did know that he was on the *Murphy* and that he was the officer of the watch when it went down later that year, but little else . . . . Thank you for sending us that poem." Another poet's daughter, Candice, reflects the emotional depth experienced by some family: "I can't tell you how moved I was to read not only your letter but the actual poem too. Every time I read them, it brings tears to my eyes. Thank you so much for your thoughtfulness." And again, an expression of surprise from a daughter Paula: "I received the copy of the *Idaho's* log and was thrilled to immediately recognize my dad's signature. Many thanks for this surprising delight!" And finally, from a son, James, who said, "Thank you for this wonderful and thoughtful gesture. You made my day."

Responding to the fact that her father had written a parody of Hamlet's third soliloquy, Jennifer said, "My mother, [*name redacted*], my sisters and I are writing because we received your letter yesterday with the poem written by my late father. . . . We wanted to express how very grateful we are that you took the time to send us his poem. Getting to hear his voice after so many years was a wonderful gift, and to top it off, we thought the whole poem was so quintessentially him - a thoughtful and off-the-cuff rendition of Hamlet's soliloquy on something so mundane as a Navy log. We will treasure it for years to come."

Two of the more unusual responses to unexpectedly receiving a father's poem were shared by these daughters. Denise said, "I am so excited to hear about this poem from my dad. I can't thank you enough and am skipping down the hall in my office suite. Which is a law firm, and that is frowned upon but I am over the moon. You have offered me a treasure that I didn't know existed." And then Kris, a singer and songwriter, offered to write a poem about her father in response to her father's poem. We included Kris's poem in the first book, *Midwatch in Verse.*

On one occasion, the arrival of the poem did more than simply surprise the recipient. It met a long-standing need. Gayle, whose father served on USS *Landsdowne* when it ferried the Japanese signatories of the documents of surrender to USS *Missouri* in 1945 and took photos of their arrival on board the ship, sent a moving response. She had received not only the poem her father wrote but also a photo that included her father taking a photo of the Japanese representatives. Gayle said, 'I spent 50 years looking for a proof of my father, [*name redacted*], on the USS Landsdowne. I checked on the internet, went to Naval Museums, etc. . . . But I never found anything in 50 years. I had all but given up. Then, out of the clear blue my brother sent me [your material]. While my father had told me he was 'on the little boat that took the Japanese to the big boat' I never really knew the details and he did not want to talk about it. So there it was . . . He was the man on top [of the bulkhead in the photo] taking the picture of the Japanese delegation after the signing. I was so excited to see him my husband surprised me with 14" X 14" blow up of the picture for Christmas. . . . Lo and behold, you enter my life with one of the greatest surprises of my entire life. Now you must realize my father died when I was only 23. I did not know him to ever do any writing, poetry or prose. You have shown me a side of him I never would have suspected were it not for you. It is an incredible thing you do. One that will never be forgotten for you have changed the memories of this life."

Kathy, the wife of one of the poets, expressed similar sentiments: "It was such a treat to receive this! Wanting to share it, I've sent copies to his family and close Navy friends. Steve would have been so touched and honored by your letter! . . . Many thanks for this wonderful memento. It's brought tears but also many smiles, wonderful memories, and a peek into Steve's life and thoughts before I met him. I so appreciate it! . . . I wasn't aware that the New Year in verse was a Navy tradition. I love it! Thank you for rounding them up and publishing them! What a treasure!"

Leslie, a daughter of one of the poets, said, "I just received your letter and copy of my dad's poetic log entry yesterday. I was surprised and delighted with this information! . . . I look forward to sharing this poem with my 95 year old mother. She does not recall him mentioning this tradition either! However, it is fitting that it arrived at nearly the 15th anniversary of his passing in 2010."

It wasn't always only the families of the poets who were surprised. A serendipitous string of events occurred when David Johnson posted the first deck log poem on his website/blog, *Casks and Quivers* (https://www.socpsy.com/casks/) in January of 2020. After posting the poem and a short biography of the poet, the poet's grandnephew Randy commented on the post, saying, "As it happens, I am now sitting here with my mother, the last remaining niece of [*name redacted*]. She is 88 and still kicking. Mom's brother, Gary, lives in Philadelphia. I had no idea of this curious, but fantastic, tradition. We excitedly enjoyed seeing his thorough description of the mid-watch that night. The history of his time in the Navy is more detail than we could ever get out of him. Like many others, he certainly did not like talking to us about the war. So this incredible, direct connection for us, imagining him sitting and penning these words, was a small joy and opportunity for us to think of him again."

Then four years later, to our surprise, a person serving on the boards of both the USS *Hornet Museum* and the USS *Hornet* veterans association, who had purchased for the museum a batch of letters from the ship from 1944, responded on the original blog post. He had read David's post and revealed that the letters he had purchased "were sent from [*the original poet*]. "When I found your article above, I thought this information would be an interesting addendum on [*poet's name redacted*]. I was also hoping to find any family, and would pass this letter on to them. But no luck so far." Dave communicated with both men, and the grandnephew of the poet was able to get possession of the letters.

David was able to find several of the surviving poets from the Cold War period. One of them, Joe, said, "Thank you for your email and tracking me down after 50+ years! I must admit that I broke down while reading and remember composing [the poem] in the Wardroom that night." Another poet, Grant, shared how he came to write the midwatch poem: "I was Senior Watch Officer in [the ship], and so assigned myself that particular watch. I knew this was my last opportunity to do the New Years verse, and really wanted to. I asked the other OODs if they minded, and, to no surprise, no one did. I spent the entire 3 days to Subic Bay writing and rewriting, because I really wanted to do it in Iambic Pentameter, since my favorite poem growing up was The Cremation Of Sam McGee, by Robert Service. . . . I am glad you are undertaking the project. I am very big on tradition, and this was one of the best." Dennis, a poet who wrote the deck log poem during the Vietnam War, said when he received his copy, "It did indeed bring back many memories. . . . as we had been in our first combat only a few days earlier, the eve of Christmas Eve. It gives you a different perspective." He then signed off, wishing us "Fair winds and following seas." Don, who also wrote his poem while serving in Viet Nam, said, "Ah yes, but your Email was delightful....am 82 now and that was over 50 years ago....so thanks much." And finally, Bob said, "Again, thank you for hunting me down and giving me the opportunity to read my entry after all these years (54 to be exact!)."

Lori, the daughter of one of the poets, was comforted during a difficult time by the words of her father's poem. "The timing of your letter could not have been better. My father died many years ago, and this year, his wife/my mother, passed away at 102. The holiday season was a little less bright, but the words "HAPPY NEW YEAR—-HAPPY NEW YEAR—Let us all make merry—" felt almost like getting a message from the great beyond. Thank you so much for your efforts in tracking us down. It is truly appreciated." And James, the son of one of the poets, expressed that his daughter would be especially excited to read the poem. "My older daughter (age 34) loves to 'rummage' through boxes at our house when she visits–looking for family history. She will absolutely love this letter. Thank you for taking the time to locate me."

Some family members were so happy to receive the poems that they gave them as special gifts to their loved ones. Larry, the son of a poet, shared his excitement. "Thanks so much for this priceless memory. I'm going to have it framed and present it to my Mom as a Christmas present. She maintains a corner of her house with mementos of my Father's service." Denise, a poet's daughter, said, "You made my day and my Mom's. Her 87th birthday was this weekend and I gave her the poem, blown up and laminated. She reports she was on the ship that night, as a guest because it was New Year's Eve. She said the meal was wonderful! She had never known about this poem and was overjoyed to have this given to her. She said you brought her a gift from God."

One of the most enjoyable things about hearing from the families of the poets was the personal stories they shared about their father/grandfather/uncle. The daughter of an Admiral told us her dad "sent my mother a number of poems during his years away. And she would respond to him with poetry also." Mary Ann, a daughter, shared childhood memories: "Daddy did have a great sense of humor. He had everyone laughing at his retirement party . . . . I have a picture of me laughing at my wedding during his toast . . . . He was a great pianist and used to play the organ at the Naval Academy. His dad played the piano for silent movies and was the organist and choir director at St. James in Cleveland for many years, so he came by his talent naturally. I loved to sit next to him on the piano bench and sing when he played from his Opera book. I still can't hear the Grand March from Aida without thinking of him. He was bigger than life. I still miss him after 35+ years." Mary Ann went on to share that her father was the ranking officer on the ship when the Pearl Harbor attack began. "He was asleep in his bunk when the bombs started. He ran up on deck in his skivvies and ordered the ship out of the harbor. I think the [the ship] was one of only two that got out. . . . Daddy decided after he stepped on a hot shell casing to take time out to get his clothes (and shoes) on."

One son, Rick, described the impact his father had on the rest of the family. "My Dad was a proud Tin Can Sailor and was medically retired as a LtCdr and then he married my Mother, a Navy nurse in WW II. All of us followed in their military service footsteps, my sister in the Air Force, me and my next two brothers in the Marine Corps and my youngest brother in the Navy submarine service." Steve, the son of another poet told us his father "liked reading western novels -- Zane Grey and similar authors. He was a good provider and faithful husband and father. Thanks for a glimpse into a part of his life he never talked about." Jack, another son, revealed an unusual fact about his father. "My father's reply to his major at the University of New Mexico would be that he majored in 'Rodeo' a favorite non-academic activity, which he was good at and loved to do."

And finally, Ed, whose father died when Ed was only four years old, said, "Wow! I just got chills reading your email! That was my dad! And from what I have learned about him, this fits him to a T! . . . I learned of some things from my mom, and some of what I know is he was very smart, clever, funny, and a big fan of Gilbert and Sullivan! In fact, he was in several plays (musicals) in college and I know for sure one was Gilbert and Sullivan's Mikado.

Thank you so much for this very welcome and surprising email! GO NAVY!"

For David, Gary, and Ken, after all the long hours researching and writing about the unique Navy tradition of doing the first deck log of the year in verse, perhaps the best response to it all is, as Ed said: GO NAVY!

# SOURCE NOTES

By far, the most used source for content of this book is the National Archives. All deck logs for US Navy ships from 1801 to 1983 are held by the Archives. Many logs are digitized and available online. A project to digitize Vietnam-era logs continues as this book is being written.

**Preface**

[1] "Mid-Watch in Verse – USS *Maryland.*" Socpsy.com. https://www.socpsy.com/casks/mid-watch-in-verse-uss-maryland/.

[2] D. E. Johnson and G. M. Guinn, *Midwatch in Verse: New Year's Deck Log Poetry of the United States Navy, 1941–1946* (Jefferson, NC: McFarland, 2023).

[3] Naval History and Heritage Command, "U.S. Ship Force Levels," https://www.history.navy.mil/research/histories/ship-histories/us-ship-force-levels.html.

**Introduction**

[1] D. E. Johnson and G. M. Guinn, *Midwatch in Verse: New Year's Deck Log Poetry of the United States Navy, 1941–1946* (Jefferson, NC: McFarland, 2023).

## Sidelight 1 - Land, Sea, and Space: Unusual Places for Poems

[1] R. W. Connell and William P. Mack, *Naval Ceremonies, Customs, and Traditions*, 6th ed. (Annapolis, MD: Naval Institute Press, 2004).

[2] "Ships Log 000001 Jan 2001," Astronautix.com, http://www.astronautix.com/s/shipslog000001jan2001.html.

## Chapter 1 - As Rhyme Goes By: Complaints About Writing Poetry

[1] The Guppy class submarines were developed following WWII. American subs were redesigned and refitted for greater underwater speed, as part of the Greater Underwater Propulsion Power Program, to meet the escalating needs of the Cold War Navy. According to John J. Roberts, "After the war the Navy obtained two German Type XXI boats. The two German Boats, the U-2513 and U-3008, were studied and tested extensively. The lessons learned: increase the battery capacity, streamline the boat's structure, add a snorkel, and add a better fire control system." The new design the Navy came up with, called the TANG Class, made the existing fleet boat obsolete, but it was so expensive that "the Navy had to find a way to upgrade the existing fleet boats to match the TANG class's ability. . . . The majority of the information about the operations of the Guppy's during the Cold War remains classified to this day."

[2] Shakespeare's iambic meter consisted of the simple use of a weak accent followed by a strong accent, as in his famous line "To be or not to be." Lt. Werenskjold's anapestic meter consists of a more complicated pattern of two weak beats followed by a strong beat, as in Clement Clarke Moore's

"'Twas the night before Christmas and all through the house."

[3] John J. Roberts, "GUPPY Submarines and the Cold War," *Quiet Warriors* (blog), accessed April 21, 2025.

**Chapter 2 - Triumph and Tragedy: The Space Race, Popular Culture, and Conflict**

[1] Cabral, Laura. "The History of Hula Hoops." *Backthen: History*, September 19, 2022.

Retrieved November 28, 2025.

**Sidelight 3 - Should old acquaintance be forgot...: A Midwatch Poet Responds**

[1] Mr. Sayers is a former naval officer who served on board a Pacific Fleet destroyer escort and in the Pentagon on the staff of the Assistant Secretary of Defense (Public Affairs). An IBM veteran, he lives in Ridgefield, Connecticut. He is a prolific author of naval ship history including the following books:

K. W. Sayers, *Uncommon Warrior: 200 Years of the Most Unusual American Naval Vessels* (Annapolis, MD: USNI Press, 2012).

K. W. Sayers, *U.S. Navy Auxiliary Vessels: A History and Directory from World War I to Today* (Jefferson, NC: McFarland, 2019).

**Chapter 3 – Wine, Women, and ........**

[1] J. A. Symonds, *Wine, Women, and Song: Medieval Latin Students' Songs* (London: Chatto and Windus, 1884).

[2] Benjamin Franklin, *Autobiography of Benjamin Franklin* (New York: Henry Holt and Company, 1916).

[3, 5] Naval History and Heritage Command, "Alcohol in the Navy, 1794–1935," August 18, 2017, https://www.history.navy.mil/research/library/online-reading-room/title-list-alphabetically/a/alcohol-in-the-navy.html.

[4] R. Barouxis, "Examining the History of Drinking Culture on Ships through Colonial-Era Grog and Madeira," *Old North Church*, July 14, 2017, https://www.oldnorth.com/blog/examining-the-history-of-drinking-culture-on-ships-through-colonial-era-grog-and-madeira/.

[6] U.S. Naval Institute Staff, "A Hundred Years Dry: The U.S. Navy's End of Alcohol at Sea," *USNI News*, July 1, 2014, https://news.usni.org/2014/07/01/hundred-years-dry-u-s-navys-end-alcohol-sea.

[7] Torpedo Bay Navy Museum, "Splice the Mainbrace," https://navymuseum.co.nz/explore/by-themes/customs-and-traditions/splice-the-mainbrace/.

[8] Naval History and Heritage Command, "Women in the U.S. Navy," July 29, 2024, https://www.history.navy.mil/browse-by-topic/diversity/women-in-the-navy.html.

[9] A. L. Smallwood, *Shore Wives: The Lives of British Naval Officers' Wives and Widows, 1750–1815* (unpublished thesis, Wright State University, 2008).

[10] D. Fee, "The Lure of the Sea: The Story of the Sailor's Valentine," *Nantucket Online*, Fall/Winter 2019, 107–113.

[11] Naval History and Heritage Command, "Enterprise Sailor Tells a Love Story," February 12, 2013, https://www.history.navy.mil/content/history/museums/nnam/education/articles/objects-of-history/enterprise-sailor-tells-a-love-story.html.

## Sidelight 4 - Ships as Women: Personal Pronouns She/Her

[1] "Why Are Ships Referred to as 'She'?" Martide.com. https://www.martide.com/en/blog/why-are-ships-referred-to-as-she.

[2] "Royal Navy Drops 'She' for Warships," *The Guardian*, April 26, 2019. https://www.theguardian.com/world/2019/apr/26/ships-she-royal-navy-language-row-female.

## Chapter 4 - Late Night Laughter: Humor on the Midwatch

[1] S. Williamson, "Comedy in Combat Culture: Understanding the Use of Humor in Crisis and Conflict," *Military Review* (September–October 2023), https://www.armyupress.army.mil/Journals/Military-Review/English-Edition-Archives/September-October-2023/Combat-Culture/.

[2] D. E. Johnson, G. M. Guinn, and K. D. Keith, "Leatherneck Bards: Marines Embrace Unusual Duty Aboard Navy Ships," *Leatherneck Magazine* (January 2025): 56–59.

[3] S. K. McBride and J. Ball, "#TheSmoreYouKnow and #emergenycute: A Conceptual Model on the Use of Humor by Science Agencies during Crisis to Create Connection, Empathy, and Compassion," *International Journal of Disaster Risk Reduction* 77 (July 2022): 102995.

[4] H. Jones, "Is Self-Deprecating Humor an Unhealthy Habit?" *Verywell Health*, October 16, 2023.

**Chapter 5 - Duck and Cover: The Threat of Annihilation in Cold War Deck Log Poems**

[1] Janet Pruitt, "How 'Duck-and-Cover' Drills Channeled America's Cold War Anxiety," *History*, March 26, 2019, updated July 17, 2025, accessed November 1, 2025, https://www.history.com/articles/duck-cover-drills-cold-war-arms-race.

[2] A. Mitchell Palmer, "The Case Against the 'Reds,'" *Forum* 63 (February 1920): 173–185. Edited and introduced by Jennifer D. Keene, Teaching American History, accessed November 3, 2025, https://teachingamericanhistory.org/document/the-case-against-the-reds/.

**Sidelight 6 - Give Me Liberty………**

[1] Naval History and Heritage Command, "Liberty (AGTR-5-III)," https://www.history.navy.mil/research/histories/ship-histories/danfs/l/liberty-agtr-5-iii.html.

[2] *Court of Inquiry: USS Liberty Incident*, http://www.gtr5.com/evidence/ncoi.pdf.

[3] "Roster of USS Liberty Including Casualty Numbers and Awards," http://assaultontheliberty.com/roster.htm.

[4] "USS Liberty," *Wikipedia*, https://en.wikipedia.org/wiki/USS_Liberty.

The National Security Agency maintains a large number of declassified documents regarding the Liberty incident. Links can be found at the following URL:

https://www.nsa.gov/Helpful-Links/NSA-FOIA/Declassification-Transparency-Initiatives/Historical-Releases/USS-Liberty/.

### Sidelight 7 - A Mind of Metal and Grace

"Charles McMahon Obituary," *The Philadelphia Inquirer*, January 19, 2023, https://www.inquirer.com/obituaries/charles-mcmahon-obituary-university-pennsylvania-metallurgy-research-20230119.html.

### Chapter 7 – Phantoms of the Deep: Hauntings and Visitations at Sea

[1] Twentieth century hauntings are occasionally reported. In his book, *The Spectral Tide: True Ghost Stories of the U.S. Navy*, Eric Mills includes a chapter about hauntings on battleships. USS Texas (BB 35) and USS North Carolina (BB 55), both museum ships, have long histories of hauntings. A ghostly spirit given the moniker "Red" appeared often on Texas usually sporting a smile. Aboard North Carolina ship tenders reported spirits pulling pranks, sometimes acting menacingly, and occasionally being seen peering through portholes.

Eric Mills, *The Spectral Tide: True Ghost Stories of the U.S. Navy* (Annapolis, MD: Naval Institute Press, 2009).

### Sidelight 8 - Admiral John Hoskins: The Man with a

**Wooden Leg and an Iron Will**

[1] *Life* magazine (archival copy), Internet Archive, https://web.archive.org/web/20160430001002/https://books.google.com/books?id=t0oEAAAAMBAJ&pg=PA72.

[2] "John Hoskins (Officer)," *Wikipedia*, https://en.wikipedia.org/wiki/John_Hoskins_(officer).

**Chapter 8 – Leatherneck Bards: Marines Embrace Unusual Duty Aboard U.S. Navy Ships.**

[1] This article from the January 2025 issue of *Leatherneck* is reprinted with the kind permission of the Marine Corps Association, the professional organization of the United States Marine Corps. https://www.mca-marines.org/.

[2] Naval History and Heritage Command, "The Corps' Salty Seadogs Have All but Come Ashore," https://www.history.navy.mil/research/library/online-reading-room/title-list-alphabetically/c/the-corps-salty-seadogs-have-all-but-come-ashore-seagoing-traditions-founder-as-new-millennium-approaches.html.

[3] Office of the Chief of Naval Operations, *OPNAVINST 3120.32C*, sec. 371.1, https://cmapspublic3.ihmc.us/rid=1H8Y4328Y-1BTNTNC-95FX/3120.32C-1.pdf.

[4] Naval History and Heritage Command, "General Order No. 99: Prohibition in the Navy," https://www.history.navy.mil/research/library/online-reading-room/title-list-alphabetically/g/general-orders/general-order-no-99-prohibition-in-the-navy.html.

## Sidelight 9 – Women in the Navy

[1] Navy League Honolulu, "Women in U.S. Navy," https://www.navyleaguehonolulu.org/women-in-us-navy.html.

[2] U.S. Navy, "Admiral Lisa Franchetti," https://www.navy.mil/Leadership/Flag-Officer-Biographies/BioDisplay/Article/3148210/admiral-lisa-franchetti/.

## Chapter 10– Midwatch Deck Log Poetry Aboard US Submarines: The Early Cold War Years

[1] This article from the December 2024 issue of *The Submarine Review* is reprinted with the kind permission of the Naval Submarine League, a non-profit organization promoting the importance of submarines and the undersea to our national defense. To join, support, or learn more, visit www.navalsubleague.org.

[2] "Submarine Poem," CRW Flags, https://www.crwflags.com/chesbase/subpoem.html.

[3] "Submarine," AllPoetry.com, https://allpoetry.com/Submarine.

[4] Siew David Hii, "Ships in the Desert and the Waterer," *Electric Literature*, https://electricliterature.com/ships-in-the-desert-and-the-waterer-by-siew-david-hii/.

## Sidelight 11 – The Long and Short of it

[1] The 1962 Currituck poem contains quite a bit of mundane information throughout its five pages. Readers who wish to see the entire poem may use the following URLs to access each of the pages in the log.

Page 1: https://tinyurl.com/yfbp9eya
Page 2: https://tinyurl.com/3m4aspym
Page 3: https://tinyurl.com/4fhmhwd7
Page 4: https://tinyurl.com/57akz4f2
Page 5: https://tinyurl.com/3zz568dd

**Sidelight 12 - From Navy Steel to Kinetic Sculpture**

[1] Princeton University, "James Seawright, Kinetic Sculptor and 'Godfather' of Creative Arts at Princeton, Dies at 85," March 1, 2022, https://www.princeton.edu/news/2022/03/01/james-seawright-kinetic-sculptor-and-godfather-creative-arts-princeton-dies-85.

**Chapter 12 – Bards of Viet Nam**

[1] Edward J. Marolda, *By Sea, Air, and Land: An Illustrated History of the U.S. Navy and the War in Southeast Asia* (Washington, DC: Naval Historical Center, 1994), xiii.

[2] Most of the historical information about the ships in this article was taken from NHHC DANFS.

[3] Eric Mills, *The Spectral Tide: True Ghost Stories of the U.S. Navy* (Annapolis, MD: Naval Institute Press, 2009); R. Winer, *Ghost Ships: True Stories of Nautical Nightmares, Hauntings, and Disasters* (New York: Berkley Books, 2000).

[4] D. E. Johnson and G. M. Guinn, *Midwatch in Verse: Deck Log Poetry of the United States Navy, 1941–1946* (Jefferson, NC: McFarland Publishing, 2023).

[5] D. E. Johnson and G. M. Guinn.

[6] Marolda, 45.

[7] Alex R. Lazalere, "The Long Blue Line: 'Tip of the Spear'—Coast Guard Joins the Fight in Viet Nam," *MyCG* (2022), retrieved August 26, 2024.

[8] Lazalere.

[9] Thomas J. Cutler, "From Blue to Green and Brown," *Naval History*, December 2015; reprinted in *The U.S. Naval Institute on Vietnam: Coastal and Riverine Warfare*, ed. Thomas J. Cutler (Annapolis, MD: Naval Institute Press, 2016), 11–12.

[10] Glen E. Helm, "Surprised at Tet: U.S. Naval Forces in Vietnam, 1968," originally published in *Pull Together* 36, no. 1 (Spring/Summer 1997): 1–5, retrieved from NHHC website August 13, 2024.

[11] Dave Curry, "Anecdotes," *Helisoma*, submitted by Kees Helder, April 19, 2015, retrieved August 13, 2024.

[12] Curry, "Anecdotes."

**Sidelight 13 – A Navy Man for Life**

[1] "Professor Wayne P. Hughes, Jr., Captain, USN (Ret.)," *Phalanx* 53, no. 1 (March 2020): 24–31.

[2] T. Vencill, "In Memoriam: Legendary Strategist and Fleet Tactics Author Wayne P. Hughes," December 11, 2019, https://nps.edu/-/in-memorium-legendary-strategist-and-fleet-tactics-author-wayne-p-hughes.

[3] Wayne P. Hughes Jr., "Naval Tactics and Their Influence on Strategy," *Naval War College Review* 39, no. 1 (1986): 2–17.

# APPENDIX A

## Oh, the Nights are Calm and the Breeze a Balm. . . Excellent Poems From the Midwatch: A Sampler

In reviewing several thousand midwatch poems, we have seen many tortured rhymes, and awkward, but game, attempts to establish a consistent poetic meter or cadence. We have been moved to laughter by funny stories woven into deck logs and touched by the human experiences included by some young OODs. They have sometimes used vocabulary that stretches a little too far in the effort to mold mundane log content into a credible poem; and they have sometimes fallen short when trying to emulate famous poets.

Yet, although most of these logs do not represent great poetry, a few stand out for qualities that make them fine poems. In this chapter, we discuss a few examples of midwatch logs that stand on their merits as good poetry. These verses are the work of sailors who, despite facing the perils of war, the challenges of bad weather, or the press of other duties, showed themselves to be clever, literate, and creative. In short, they are good writers.

One such poet, Lt. j.g. C. A. Winters, shows a flair for verse in his log entry for January 1, 1956, aboard the USS *Gudgeon* (SS 567). Commissioned in 1952, *Gudgeon* was the first American submarine to circumnavigate the earth, and although sub logs frequently did not disclose their ship's location or course, in this case Winters was free to wax poetic about his ship's destination, with an echo of Robert Service's *The Cremation of Sam McGee*. Like Service, Winters writes mainly in anapestic tetrameter, with a consistent rhythm and enough variety to keep the poem interesting and moving along, and he skillfully maintains both line-ending rhymes and internal rhymes within lines:

Oh, The night's as calm as a desert palm
        In the old South China Sea
In a shipping route just five days out
        Of Hong Kong, B.C.C.
Manila scenes in the Philippines
        Are ahead in the rising sun;
And the C.T.G. of this jamboree
        Is Commander, CarDiv One.
The helmsman stands, helm in hands,
        Steering 160 true
And maneuvering bells show a speed that tells
        Of a placid cruise on the blue.
"Six knots ahead," The Captain said;
        "Six knots," said the OOD,
"And the amps the screws don't need to use
        We'll put in the battery".
Just take it slow; No place to go
        But the rendezvous point next.
Let the surface fleet in formation neat
        Run around like a hound dog hexed.
The big Kearsarge and the Admirals barge
        (Bennington, C.V.A.)
Roar this way and that at the drop of a hat;
        Six small boys lead the way.
But the days go slow and there's more to go
        'Ere the exercise is done
So we sail with ease and enjoy the breeze
        While the task group lives on the run.
Oh, the nights are calm and the breeze a balm

In the old South China Sea,
On a shipping route just five days out
Of Hong Kong, B.C.C.

Another good poet welcomes 1969 aboard USS *Snohomish County* (LST 1126) as she steams for Danang, Republic of Vietnam. Although Lt. j.g. J.S. Huntsberry does not divide his poem into stanzas, he nevertheless maintains the flow characteristic of ballad stanzas by rhyming alternating (e.g., second and fourth) lines:

Above my head, the mast, an inverted pendulum,
is a metronome against the stars.
Our bark of steel bucks and rolls, pursuing
foamy wave, capped by Mars.
The clock's minute hand moves ahead two, then back one,
ahead two, then back one. . .
While a lemon pathway is offered by the lunar orb,
ancient antagonist of the sun.
The compass ticks in monotonous rhythm with the
droning throb of engines ahead one third (5 kts).
To this changeless tune dance memories of the year past,
broken occasionally by the helmsmen's word . . .
"Steady course of one three zero," his notes seem
to blend with the overall harmony . . .
Of the night, the stars, and a ship, sailing
the sea.

The *Snohomish County* was a World War II-era tank landing ship that received eight battle stars for her service in Vietnam. Huntsberry's poem was her penultimate midwatch verse; she would sail on only one more New Year's Day, on her way to being decommissioned in 1970 and scrapped in 1971.

At the dawn of 1957, USS *Seagull* (MSC(O) 55) OOD Lt. j.g. J. R. Talbot, seems to try to replicate the meter of the old folk song "Oh Where, Oh Where Has My Little Dog Gone?" but with an extra beat or two at the end of each line:

> Oh where are the girls with the raven-dark hair,
>         And where is the cup of good cheer?
> They're all on the beach here in old Norfolk-town,
>         With the liberty boys we do fear,
> While we keep the ship at the bulkhead between
>         The fourth and the fifth chilly pier
> At Norfolk Shipbuilding—their Berkley plant here—
>         At the start of this lonely new year.

The *Seagull*, a World War II minesweeper originally classified as YMS 402, was reclassified AMS 55 and given her name in 1947. She served in the Atlantic and Caribbean during the Korean conflict and then was reclassified MSC(O) in 1955. Talbot's poem would be her last New Year midwatch log— She was decommissioned later in 1957.

USS *Taluga* (AO 62) was a durable oiler that received four battle stars in World War II, four more during the Korean War, and six campaign stars during the war in Vietnam. Her OOD on January 1, 1962, Lt. F. B. Harris, shows pride in his ship and its crew in these stanzas from the midwatch log:

> Come, all ye jolly sailors bold
> While I sing a song of good black gold
> As an Oiler's story I unfold—
> Huzzah for the TALUGA!
>
> We steam 'round the world
> With our ensign unfurled
> As twenty-four hours a day we do toil
> To deliver the cargo of good rich black oil
> That's on board of the good ship TALUGA . . .

We spent Christmas Day
At sea, underway
Ten days from Pearl Harbor, our last port of call;
The following morning we made a landfall
From the bridge of the good ship TALUGA . . .

Now let each fill his glass
To his favorite lass;
A health to the Captain and Officers true
And all that belong to the jovial crew
On board of the good ship TALUGA.

So come, ye jolly sailors bold
As this Naval Custom we uphold
And sing out clear, a thousand-fold:
Huzzah for the TALUGA!

Lt. Harris maintains a consistent meter in his poem, using a pair of rhyming couplets followed by a repeated return to TALUGA in each stanza. Exceptions are the first and last stanzas, where he uses a rhyming triplet, again followed by the repeated exclamation: "Huzzah to the TALUGA!" He may well have been busy with more pressing tasks, but he clearly gave careful thought to the structure of his midwatch verse.

On January 1, 1967, USS *Serrano* (AGS 24) was moored at Pearl Harbor. Her OOD for the midwatch, Lt. j.g. R. D. Butterbrodt combines a heady Shakespearean air with a sly sense of humor as he begins his log entry:

Now brings the gentle muse, poor goddess
unfamiliar in these bounds of tempered steel,
her phantom power succor to my suit
as fitfully I ply the leaves of Webster's tome;
not modern knight nor venerable Bowditch
delved this heady problem, verse to spin,
which time, flitting at this present eve

from year to year, leaves little chance
for thoughtful care—The breeze is in the shrouds,

alight with strands of multicolored globes,
and round about more units of PACFLEET
are draped with trappings of the season.

"Venerable Bowditch" refers to *The American Practical Navigator,* a nautical guide authored by Nathaniel Bowditch and first published in 1802. The *Navigator* has been revised over the years and remains in use today. And, of course, "Webster's" refers to the dictionary, as he seeks to fulfill the unfamiliar role of midwatch poet. *Serrano* was commissioned as a tug in 1944, then taken out of service in 1950, only to come back ten years later, converted to a survey ship.

The USS *Enterprise* (CVAN 65) was a groundbreaking ship. She was not only the world's first nuclear-powered aircraft carrier, but also the longest naval vessel ever built (342 meters). Nicknamed the "Big E," *Enterprise* inspired pride in her crew—pride reflected in this midwatch log, penned by Lt. R. R. Conger on January 1, 1964:

> The atom that split during World War II has changed the ships, vast oceans knew.
>
> In sixty-one a hull was done, to be the first atomic run. Name her big, think of her size, thoughts all A'jig, let us surmise.
>
> Her name wouldn't be of the rank and file. But something grand to fit her style. Something Historic; of red glared skies; of America's freedom and ENTERPRISE.
>
> We'll call her that, just as of yore, for seven great ships who've gone before. What's really the meaning

behind that name? American's freedom; our nations fame. A spirit that's willing and a crew that tries Thats all we need for ENTERPRISE.

They launched her wide, they launched her deep, and charged her with our seas to keep. They gave her men, they gave her planes, and sent her out on the ocean's lanes. . .

Know ye then, ye men of red. We'll make this ship a thing of dread. We're even alert and unafraid of missiles launched on a mighty raid. By acts and deeds you jeopardize and teach the wary to despise.

So boast and jest and threat to bury. On little things we do not Tarry, but neglect your God and risk a fight and then you'll trigger ENTERPRISE might.

We're made of men, not just of steel. With strong, straight hearts, in truth we feel our greatest goal, over destination is that of serving a righteous nation. We'll keep her free, with bright blue skies, with blood, with tears, with ENTERPRISE.

Lt. Conger uses a series of anapestic feet in each line to maintain not only the rhythm of the poem, but also an energetic feel. In addition to extolling the virtues of his ship, he also acknowledges the Cold War in his reference to the "men of red," and gives a nod to the Navy's "seven great ships" that have previously carried the name *Enterprise*. *Enterprise* would eventually serve for fifty-one years, including the Vietnam and Gulf wars.

Chief Electricians Mate T. L. Hendrickson of destroyer tender USS *Sierra* (AD 18) reveals more than his electrician's skills in the ship's midwatch log of January 1, 1970. He invokes not only Janus, the Roman god of beginnings whose two faces allow him to look both forward and back, but also

Neptune, Roman god of the sea; Hephaestus, Greek god of fire and metalworking; and Aries, the Greek god of war. In this opening stanza, we see the image of Janus surveying both the past and things to come:

> Through the clouds overhead a rare visage appears,
> 'Tis Janus, who heralds the breaking new year.
> One head sees the future, one head sees the past,
> The porter of heaven hails us and asks:
> "Who? Whence? Whereto?"

Hendrickson follows up with additional references to Greek and Roman mythology, as well as *Sierra*'s job of servicing and repairing other ships, as he tells us more about his ship's status and location. And he does it all while maintaining a consistent meter and rhyme scheme:

> We, ancient god, are men of the sea;
> We sail on Sierra that the world might be free.
> The trident of Neptune emboldens our shield,
> the tools of repair our strong arms always wield.
>
> This night finds us moored to Virginia's shores,
> In the city of Norfolk: Ever far from the war?
> Six lines are all tripled to berth two one one,
> we're starboard side to at pier twenty-one.
>
> Destroyers and submarines call these piers home,
> a haven, a shelter, from the sea's raging foam.
> The New and the Steinaker nest by our side,
> together we rise on the incoming tide.
>
> Haephestus may rest now and not be offended,
> our boilers and dynamos need not be tended.
> Rest nephew of Neptune, Haephestus oft-chided,
> our heat and our power from the pier are provided.

You see, ancient Janus, these other ships here?
We all work together, preserve freedoms dear.
United States Navy our title, our fame,
Atlantic fleet units, we play Aries' game.

ASWFORLANT's commander is near,
He's SOPA tonight; we wish him good cheer.
May we pass now, old porter? The old year is done.
"But where are you going? From where have you come?

For twenty-five years Sierra has labored
to tend fighting ships and send them all seaward.
For twenty-five more she'll work if she can,
this old gray ship with the helping hands.

Commissioned in March of 1944, *Sierra* spent the summer providing service to ships at Pearl Harbor, eventually making repairs for ships preparing for the attack on Iwo Jima. Following World War II, she called a number of ports home, and for some years served as flagship for Cruiser Destroyer Flotilla 4. Following refurbishment in the early 1960s, Sierra went on to serve in a variety of ways, including her crew's work helping the cleanup following Hurricane Frederic in 1979. At that time, *Sierra* underwent another overhaul, this time including installation of quarters for her first female officers. The ship's long service included aid in the aftermath of another hurricane, Andrew, in 1992. After nearly a half century that spanned the globe and a wide variety of forms of service, USS *Sierra* was decommissioned and scrapped in 1993, more than twenty years after Chief Petty Officer Hendrickson's midwatch log.

Another reference to Greek mythology appears in the midwatch log of USS *Ajax* (AR 6) at the beginning of 1964. Master Chief Electronic Technician. F. Riefer ties his ship's name to the mythical Greek hero, as he writes the log in the voice of *Ajax*. Although he alludes to the Greek hero as a

Spartan, Ajax, according to legend, was actually from the island of Salamis. That error is, of course, forgivable, and use of the word Spartan probably makes for a better poem.

> I'm just out of the yards—with a fresh coat of gray,
> My proud two stars—fly high o'er the bay.
> My booms are varnished—I have bright shiny brass,
> I'm ServPac's finest—the best of my class.
>
> I'm AR Number Six—Ajax is my name,
> After that speedy Spartan—the very same.
> For 20 long years—I've served the fleet,
> My "customers" will tell you—I'm hard to beat.

Riefer goes on to provide a brief lesson in Japanese culture:

> I'm moored in Japan—in Sasebo Ko,
> The "Ko" means harbor—in case you didn't know.
> I'm in India Basin—Berth Number Eight,
> They like me here—I'm near the main gate.
> I've been in Japan—for so many a year,
> My crew thinks "Asahi"—is "King of all beer."
> It's been so long—you may think I'm "whacky,"
> But instead of black oil—I prefer saki.
>
> I even watch "Sumo"—Taiho is best,
> He's to the East—as Staubach is West.
> When they sent me here—to promote good will,
> They realized the "Spartan"—could fill the bill.
>
> I'm learning their language—"Wakarimash 'ta ka?"
> Which means "Do you get it?"—'Sorta easy, eh?"
> Or how about this one—"Oyas' mi-nasai?"
> That means good-night—"Sayonara's" good-bye.

After spending several stanzas describing the ship's mooring, sources of power and water, and other nearby vessels, Riefer, like nearly every New Year OOD, ends with wishes for a happy new year:

Well that ends my story—it's now '64,
None of us will ever—see '63 anymore.
So please accept from AJAX—and also from me,
"Happy New Year to all"—who follow the sea . . .

The *Ajax* poem has an interesting structure, with the use of dashes in each line as an aid to the rhythm of the language. The rhyme scheme is a common one—AABB within each stanza, with the endings of the first and second lines rhyming, as do the last words of the third and fourth lines. Taiho in the poem refers to Taihō Kōki, a sumo wrestling star of the 1960s, and Staubach is Annapolis graduate Roger Staubach, an American football star of the era.

Like *Enterprise* and *Sierra*, *Ajax* was another long-lived ship, serving from 1943-1986, then eventually being sold for scrap. She was a repair ship, serving first in World War II. Then, after carrying out towing, repair, and salvage work for the 1946 Bikini Atoll nuclear bomb tests, she would go on to make many trips to Japan. For a short time in the early 1950s, *Ajax* provided support to ships taking refugees out of North Vietnam and returned to Vietnam in 1968 to provide repair services in the Mekong Delta. She was back in Japan in the 1960s and served as flagship of ServRon 3 in the Far East. For service in World War II, Korea, and Vietnam, Ajax received numerous service medals and commendations.

These poems reveal that their authors were not only sailors, serving their country in far-flung places, but also thoughtful, literate, and creative individuals who happily honored the midwatch verse tradition. They demonstrate awareness and knowledge of literature and culture, and they have an ear for the language. Whether on repair ships, oilers, tenders, or a massive nuclear-powered carrier, they put aside their workaday obligations long enough to leave us evidence of their good-natured humanity and indomitable spirit.

In his famous poem, Tennyson wrote:

> Break, break, break,
> On the cold gray stones, O Sea!
> And I would that my tongue could utter
> The thoughts that arise in me.

These midwatch poets give voice not only to their thoughts, but also to the reality of life aboard ships of the U.S. Navy, from Korea, to the Cold War, and on to Vietnam.

*For More News About David E. Johnson, Gary M. Guinn, and Kenneth D. Keith, Signup For Our Newsletter:*

**http://wbp.bz/newsletter**

*Word-of-mouth is critical to an author's long-term success. If you appreciated this book please leave a review on the Amazon sales page:*

**https://wbp.bz/watchstandsr**

www.ingramcontent.com/pod-product-compliance
Lightning Source LLC
LaVergne TN
LVHW010320220926
848484LV00004BB/816